Equity in Mathematics Education

To our daughters
Kate and Sarah

Equity in Mathematics Education:

Influences of Feminism and Culture

Edited by

Pat Rogers and Gabriele Kaiser

Routledge
Taylor & Francis Group
LONDON AND NEW YORK

First published in 1995
By RoutledgeFalmer

Published 2004 by Routledge
2 Park Square, Milton Park, Abingdon, Oxfordshire OX14 4RN
711 Third Avenue, New York, NY 10017

First issued in hardback 2017

Routledge is an imprint of the Taylor and Francis Group, an informa business

A catalogue record for this book is available from the British Library

Library of Congress Cataloging-in-Publication Data are available on request

ISBN 13: 978-1-138-18048-2 (hbk)
ISBN 13: 978-0-7507-0401-4 (pbk)

Jacket design by Caroline Archer

Typeset in 10/12pt Bembo by
Solidus (Bristol) Limited

Contents

Acknowledgments

We would like to acknowledge the excellent work of Steven Katz who assisted us with the editing of this book. His careful checking of references and his painstaking attention to the detail and form of individual chapters were invaluable to us and to the authors.

We wish to thank Kluwer Academic Publishers to reproduce Chapter 24 of this book which appeared first in *Educational Studies in Mathematics*, 28, 3, 1995.

Chapter 1

Introduction: Equity in Mathematics Education

Gabriele Kaiser and Pat Rogers

Research and intervention over the past three decades have greatly increased our understanding of the relationship between gender and participation in mathematics education. Research, most of it quantitative, has taught us that gender differences in mathematics achievement and participation are not due to biology, but to complex interactions among social and cultural factors, societal expectations, personal belief systems and confidence levels. Intervention to alter the impact of these interactions has proved successful, at least in the short term. Typically, interventions sought to remedy perceived 'deficits' in women's attitudes and/or aptitudes in mathematics by means of 'special programmes' and 'experimental treatments'. But recent advances in scholarship regarding the teaching and learning of mathematics have brought new insights. Current research, profoundly influenced by feminist thought and methods of enquiry, has established how a fuller understanding of the nature of mathematics as a discipline, and different, more inclusive instructional practices can remove traditional obstacles that have thwarted the success of women in this important field. Some argue that practices arising out of contemporary analysis will improve the study of mathematics for *all* students, male and female alike.

This book provides teachers, educators and other interested readers with an overview of the most recent developments and changes in the field of gender and mathematics education. Many of the chapters in this volume arose out of sessions on 'Gender and Mathematics Education' organized by the editors for IOWME (International Organization for Women and Mathematics Education) as part of ICME-7 (the Seventh International Congress on Mathematical Education) held in Québec City, Canada in August 1992. We are fortunate to have in one volume the perspectives of internationally renowned researchers and practitioners from all over the world, and from a variety of ethnic and cultural backgrounds.

However, the book does more than provide a review of current thinking in the area for we have grounded our overview in a model for understanding how change occurs. This model, developed by Peggy McIntosh (1983), arose out of an examination of the evolution of efforts in North America to loosen curriculum from a male-dominated, Eurocentric world view and to evolve a more inclusive curriculum to which all may have access. We acknowledge the danger and the difficulties inherent in attempting to classify anything into what might be narrow or restrictive categories. Nonetheless, we believe that applying this model to mathematics education provides new insight and guidance for future endeavours. Using the model as a lens for examining attempts to change the relationship between gender and mathematics education, we are able to discern and question underlying assumptions, to appreciate where we have been and to understand how certain feminist theories and cultural influences have affected and transformed our practical efforts to implement changes suggested by the research. We are then able to ask the important questions: where are we heading and to what end?

In this introductory chapter, we describe the McIntosh model and locate work in the area of gender reform of mathematics education in phases of her model (for an earlier application of the McIntosh model to mathematics education, see Countryman, 1992). In terms of the McIntosh model, we are in a transitional stage, between Phase Three (seeing women as victims or as problems in mathematics) and Phase Four (seeing women as central to the development of mathematics). This book is organized to lead the reader through this transition; as well it presents some critical perspectives and insight from researchers in developing countries which broaden and enrich the discussion.

Understanding Curriculum Reform

According to Peggy McIntosh, her typology of interactive phases of personal and curricular revision derive from her work in helping 'traditionally trained white faculty members to bring into the liberal arts curriculum new materials and perspectives from women's studies' (McIntosh, 1989). The model comprises five stages of awareness which, according to McIntosh, are patterns of realization or frames of mind which occur in succession as individual scholars re-examine the assumptions and grounding of their discipline and enlarge their understanding of the field. For example, in the field of history,

> [t]raditionally trained white ... [historians] ... were likely to move from thinking and teaching in Phase One: Womanless History, to Phase Two: Women in History. Then followed an expansion into Women as a Problem, Anomaly, Absence or Victim in History. All of these are, or can be, conceptually male-centred. I identify as Phase Four the far more daring Women's Lives *As* History, and looked toward Phase Five: History Redefined and Reconstructed to Include Us All. (McIntosh, 1989)

The issue for the authors of the chapters in this book is how to reform mathematics education to 'include us all', or more specifically, to include a greater proportion of women. Efforts in the west to achieve gender balance in mathematics education have derived from the essentially monocultural view that a certain measure of familiarity and competence with mathematics is important to every individual's future growth and economical survival. According to McIntosh (1989), those who 'think monoculturally about others, often imagine ... that others' lives must be constituted of "issues" or "problems" ...'. Thus, in attempting to change a perceived gender imbalance in mathematics participation, we have tended to see women who do not embrace mathematics as deprived, as having 'a problem' which we need to fix. When we enlarge the scope of our reform efforts, another danger of monocultural thinking is evident when we blindly apply methods that have been successful in one culture or one ethnic group to another without any, or at best with only minimal, adaptation to local needs and circumstances (see in particular, Chapter 14 by Saleha Naghmi Habibullah, for a spirited attack of this view).

Applying McIntosh's model to mathematics, we discern five phases, which we name (adapted from Countryman, 1992, p. 84):

- Phase One: Womanless mathematics;
- Phase Two: Women in mathematics;
- Phase Three: Women as a problem in mathematics;
- Phase Four: Women as central to mathematics; and
- Phase Five: Mathematics reconstructed.

This book is organized according to these phases. In this introductory chapter we draw extensively on McIntosh (1989) to describe the first three (essentially monocultural) phases of the model in the context of reform in mathematics education. However, we should warn the reader, as does McIntosh herself, that these phases do not always occur in the succession given here. Individuals may weave back and forth between and among the phases. Part 1 provides an overview of approaches to gender reform of mathematics education which challenge the monocultural assumptions and deficit philosophy of Phase Three and by virtue of their debt to feminist influences exhibit qualities which place them somewhere in the transition between Phase Three and Four. It is also important to remember that the McIntosh model has grown out of a North American perspective. The situation may be very different in other countries and in particular in the developing world, an issue which is raised in Part 2 of the book. Furthermore, not all western countries are currently at the same stage of awareness of the issues. For example, scholars in Germany and Sweden began to work in this area only in the late 1980s (see, for example, Kaiser-Messmer, 1994 and Grevholm, this volume, Chapter 6), while work in England, Australia and North America has been in progress for over two decades. The chapters in Part 3 describe recent advances in developing a Phase Four approach to mathematics education, one in which women's experience is central to the discipline and to its pedagogy. The chapter we have used as our epilogue allows us to imagine where we will be when our work is done.

The Early Phases of Gender Reform in Mathematics Education

Phase One: Womenless Mathematics

When the editors of this book were in secondary school, it was unusual for women to specialize in mathematics beyond the compulsory years of schooling, even rarer for them to go on to university to study mathematics. None of the theorems listed in our mathematics textbooks were named after women, or if they were, this fact was not made apparent to us. The language of instruction was unselfconsciously male, examples dealt with male experience. One of us, Rogers, developed a habit of personifying mathematical objects and was not even aware that she did so until years later when one of her students pointed out that she referred to mathematical terms and objects exclusively as 'him' or 'he'. Before leaving school to enrol in a prestigious university to study mathematics, she did not think it incongruous that the book prize she received for excellence in mathematics was entitled *Men of Mathematics* (Bell, 1937). Mathematics was what men did. We learned that women had not been necessary to the development of mathematics, and were unlikely to be essential to its future development. In this phase, many women who nevertheless pursue mathematics experience silence and exclusion, a feeling that 'this is not about me', a feeling not unlike the first stage of knowing described by Belenky *et al.* (1986) and discussed by Becker in Chapter 20.

Phase Two: Women in Mathematics

This phase in mathematics education reform began in North America in the 1970s when scholars began to investigate the lives and works of the few exceptional women throughout history who had been successful in mathematics, for example, Hypatia, Sonya Kovalevskya, and Emmy Noether. Although this phase challenges the all-male face of success in mathematics, it presents no challenge to the terms of success. Despite its roots in early North African civilizations, 'Mathematics remains what white men do, along with a few token women' (Countryman, 1992, p. 77). The history of exceptional women in mathematics was injected into our experience (see, for example, Lynn Osen's 1974 reply to E.T. Bell, *Women of Mathematics*). The problem with the Phase Two — 'famous few' approach (McIntosh, 1989) is that it teaches about women mathematicians as exceptions and can convey the impression that most women do nothing of any value in mathematics, and that they can only be worthy of notice if they become more like men. It ascribes to women in the field a 'loner' status that makes them vulnerable to every setback. It leaves women mathematicians believing that they can make it on their own by virtue of sheer hard work, and that if they don't, their failure has something to do with personal merit rather than with the way mathematics culture is organized. In contrast, those who are successful may begin to fear they will not be seen as female anymore. For example, Rogers, who went to an all-girls secondary school, learned to be proud of being good at something in which women didn't excel — she was one of only three girls who continued to study mathematics to the end of secondary school and the only one who ultimately pursued a career in pure mathematics. But privately, she also began to experience a nagging feeling that perhaps she wasn't a *real* women. Her self-image during this period of her life was that of a long bean-pole with a large pulsating sphere on the top. Her internal struggle between the life of a woman and the life of the mind made it difficult for her to embrace the intellectual life she enjoyed and retain an integrated sense of herself at the same time.

A further problem with this 'woman mathematician as exception' phase is that it focuses on individual success, on 'winners and losers', and thus devalues those who prefer collaborative approaches. Group work is thus not encouraged in the classroom and mathematics is seen as the property of experts. Again referring to Belenky *et al.*, this is a phase that encourages and promotes a received view of knowledge (see Becker, Chapter 20).

Phase Three: Women as a Problem in Mathematics

An illustration of a shift in approach to Phase Three is provided by the residential mathematics camps for 15 year-old girls, entitled 'Real Women Don't Do Math!' (Rogers, 1985), which Rogers organized in the summers of 1985 and 1986. The title of this camp was inspired by the phrase 'real men don't eat quiche', popularly used at that time to challenge conventional notions of what it meant to be a man. In the playful title of Rogers' camp, we can discern the shift from seeing the woman mathematician as a 'loner' to viewing mathematics as for *all* women. We can also sense the frustration with the prevalent assumption that mathematics is a field in which women have difficulty (this is perhaps more evident in the title of the video that grew out of the camp: 'Real Women Don't Do Math! — Or Do They?').

In North America since the early 1980s, and in other English-speaking western countries such as Britain and Australia, gender reform of mathematics has been dominated by intervention projects. The largest and most influential of these projects is the North American EQUALS project which has developed a variety of teaching materials, workshops and one-day careers conferences (with titles like 'Expanding Your Horizons') with the explicit purpose of increasing women's interest in mathematics. The aim was to make female students more aware of the need for mathematics in an increasingly technologically driven world and of the importance of keeping their career options open in order to compete successfully in it, and to make their teachers and parents aware of the problem of women's poor participation in mathematics (see, for example, Kaseberg *et al.*, 1980).

Undoubtedly programmes such as these have had a crucial influence on the evolution of our thinking on the issue and they have met with much success in achieving their goals. However, as mentioned earlier, a difficulty with these approaches is that they derive from a monocultural perspective that does not notice that women may have made conscious choices to avoid a subject that was in itself alienating. Rather, it is assumed that women avoid mathematics because of ignorance concerning the importance of mathematics to their futures and the dire consequences should they avoid mathematics. A different, still popular means of intervening, is to focus on *math anxiety*, a term invented by Sheila Tobias (1978, 1994) to describe a psychological fear or anxiety associated with engaging in mathematical activity. This approach uses essentially clinical means to help women overcome their anxiety towards, and hence their avoidance of, mathematics. It too, although for different reasons, places the blame for lack of participation in mathematics firmly on the shoulders of women themselves. In all of these approaches, the mathematics itself is not questioned, only the learners. It is assumed that women have to come to terms with their problems (their ignorance of consequences, their faulty beliefs, their 'mathophobia'), and that when they do all will be well.

The views of proponents of such intervention programmes are typical of liberal feminism (see Leder, Chapter 13), or feminism of equality (see Mura, Chapter 19), in that they ' "work within the system," attempting only to improve the lot of women within a society which is otherwise left unchanged' (Damarin, 1994). It is evident in this phase, that while the focus is on important issues, such as sexism and oppression, in attempting to make mathematics a more open discipline the emphasis is on disempowerment (why women can't do mathematics, why they avoid it) rather than on empowerment (how women can learn the skills to challenge the discipline and change the mathematics). Programmes of remediation, self-help and career information do not fundamentally challenge the power of authorities. In fact they may even serve to maintain the status quo.

Gender and Mathematics Education: The State of the Art

The first part of this book provides a general overview of the approaches that have been developed during the last decade to increase women's participation in mathematics. Because of their focus on changing women, there is a temptation to identify these approaches with Phase Three in the McIntosh model. Yet, they have all been influenced by an awareness of the systemic reasons for women's under-representation in mathematical fields and for this reason, we ascribe them a transitional status between Phase Three and Phase Four (see below).

A Transitional Phase

The chapters in Part 1 of this collection exhibit a worldwide shift of the debate on the relationship between mathematics and gender, one that is profoundly influenced by feminist scholarship and by research on the cultural dependency of approaches, and indeed of performance and participation in mathematics itself. This can perhaps be seen most clearly in the evolution of the SummerMath Program described by Charlene and James Morrow. This summer camp for girls was originally conceived as an intervention programme seeking to encourage girls to do more mathematics. Over the years, however, influenced by feminist theory, in particular *Women's Ways of Knowing* (Belenky *et al.*, 1986), and by constructivist approaches to the learning of mathematics, the programme has changed its theoretical orientation. Now in this camp there is an increased emphasis on encouraging and supporting connected knowing and different ways of teaching mathematics which conform to women's preferred learning styles.

Part 2 provides a cultural perspective on the approaches in Part 1. These studies question the assumption that everyone must achieve a certain level of mathematical competence, and raise doubts as to whether models developed in western countries to achieve gender equity will be effective in all countries, or even, for that matter, with powerless groups within western countries. They point rather to the necessity of developing individual cultural-specific approaches. For example, Sharleen Forbes, in analysing the situation of Maori girls in New Zealand, reports that strategies which have been successful with girls of European origin have had no positive impact on indigenous populations. She argues instead for strategies based on an intimate knowledge of Maori culture. Other contributions in this part of the book are influenced by the culture dependence of gender differences. For example, Paul Brandon, Cathie Jordan and Terry Ann Higa, citing socio-cultural causes, point out that girls from certain immigrant groups in Hawai'i achieve significantly better levels of mathematical attainment than boys. The chapters in this part of the book illuminate the results of SIMS (Second International Mathematics Study), which show in particular that differences in mathematical attainment between countries are larger than those between the sexes (see, for example, Hanna, 1989). The chapters by Françoise Delon and Pat Hiddleston, among others, reflect another orientation within this part of the book: the effects of single-sex education. They come to the conclusion that sex-segregated education has been crucial in promoting girls' mathematics achievement.

The spirit of the transition from Phase Three and the promise of Phase Four is captured by the following quote from David Henderson, a white US mathematician:

> Recently, I was thinking back over the times that my perception of mathematics had been changed by the insights or questioning of a person in [one of my courses]. Suddenly, I realized that in almost all of those cases the other person was a woman or from a different culture than my own.... [W]hen I listen to how other people view mathematics my understanding of mathematics changes. I am certain that as women, and members of the working class and other cultures, participate more and more in the established mathematics, our societal conceptions of mathematics will change and our ways of perceiving our universe will expand. This will be liberating to us all. (Henderson, 1981, p. 13)

In the next section, we describe the fourth phase in the McIntosh model and suggest how the contributions in Part 3 of this book elucidate the process of change in this phase.

Phase Four: Women as Central to Mathematics

In this phase, women's experience and women's pursuits are made central to the development of mathematics. Work in this phase seeks to uncover privilege and to redistribute power. It emphasizes cooperation over winning and losing, difference and multiplicity over 'one right way'. Fundamentally, Phase Four shifts the 'blame' away from women by seeking to change the system, not the women.

The chapters in Part 3 anticipate this phase and are grouped into two categories according to where they place their primary emphasis: changing the pedagogy or changing the discipline. Influenced by feminism and feminist pedagogy, the chapters of the first group call for a fundamental change in pedagogical processes, one that considers the experiences of women as central to their mathematics development, and in which emotion and reason play balanced roles. Fundamental to these approaches is the theory developed by Gilligan (1982) which describes the 'different voice' of women, so frequently silenced in academia. Influential, too, is the approach to teaching developed by Belenky *et al.* (1986) in *Women's Ways of Knowing*, based on their categorization of the stages in which women come to know. Several chapters (see, for example, Joanne Rossi Becker and Pat Rogers) describe how to apply feminist pedagogy to mathematics teaching.

The approaches in the second group all question the discipline of mathematics but depart from a variety of positions. Referring to the feminist critique of the nature of science, Leone Burton's focus is developing a new epistemology of mathematics, one which does justice to women. Betty Johnston reflects on the interaction between mathematics and society, and especially on how the quantification of society both constructs and regulates our lives as women. Marjolijn Witte proposes a view of mathematics, based on constructivism, which might be more conducive to women's learning preferences.

Within the approaches covered in this part of the book, it is evident how much the discussion of gender equity in mathematics education has changed in the last decade. As already mentioned at the beginning of this chapter, Pat Rogers started her work in the area of gender and mathematics with the organization of mathematics camps for girls. At this time, it was widely believed that if we changed the conditions in which students learned, and the nature of their experience with mathematics, then the relation of girls to mathematics would automatically improve. During the development of her work in this area, it became clear to Rogers that mere system-tinkering was not adequate to the problem at hand, and this is reflected in the subtitle of her (1985) paper, 'Real women don't do math — With good reason!'. Since then, she has developed an approach that fundamentally changed her own mathematics teaching, her relationship to her students, and the learning experience she provides. A similar shift can be observed within the change of the position of Leone Burton from her approach in the 1980s (see Burton, 1986) to her questioning of mathematics epistemology in this book.

In this context, it is also interesting to note that alongside debates about specifically female ways of knowing, fundamental changes in research methodology have also taken place. Most of the studies that were conducted during the Phase Three period of reform were designed to describe differences between boys and girls in achievement, or differences in their attitudes towards mathematics. Generally speaking, they were based on large student samples and the results were usually analysed using traditional quantitative methods. Representative of this kind of research is Armstrong's analyses as part of the NAEP (National Assessment of Education Progress) (Armstrong, 1981, 1985). In the course of developing an emphasis on women's ways of knowing and connected teaching or feminist pedagogy, qualitative methods, which have their root in case studies, have become more and more important. Elizabeth Fennema (1994) in asserting the importance of continuing to sustain 'some research which utilizes a positivist perspective', further stated, 'I also believe that an understanding of gender and mathematics derived [only] from studies done from [a positivist] perspective will be limited. We will not deepen our understanding of gender and mathematics until scholarly efforts conducted in a positivist framework are complemented with scholarly efforts that utilize other perspectives [for example, from cognitive science and feminism], many of which are currently being utilized in mainstream education'.

Conclusion: Mathematics Reconstructed

In conclusion, we ask what might mathematics be when it is reconstructed to include us all? In Phase Five (Mathematics reconstructed), cooperation and competitiveness are in balance and mathematics will be what people do. McIntosh sees 'the work towards Phase Five taking one hundred years because it involves a reconstruction of consciousness, perception and behaviour.' For Countryman (1992), 'The words to describe the fifth phase are more elusive. I find it hard to say what the transformed mathematics curriculum will be, or how we will achieve it. Surely in this phase mathematics will help us see ourselves and others in connection with the world', (p. 75). Surely too, it will involve a fundamental shift in what we value in mathematics, in how we teach it, in how mathematics is used, and in the relationship of mathematics to the world around us.

We have used the words of Nancy Shelley to describe the final phase in mathematics education reform. In the epilogue, she questions the disciplines of mathematics and mathematics education. Departing from a traditional academic style of writing, her form itself as much a part of her argument as the arguments she summons, she stimulates us to imagine a mathematics not dominated by authorities. Shelley touches on every issue raised in this book. In questioning monocultural views of mathematics and mathematics education, and in examining the epistemological status of fundamental issues of mathematics and mathematics teaching, she develops a vision of another type of mathematics. We leave it to our readers to say whether this is a vision of a mathematics close to that Henderson (above) described as 'liberating to us all'.

References

ARMSTRONG, J.M. (1981) 'Achievement and participation of women in mathematics: Results of two national surveys', *Journal for Research in Mathematics Education*, 12, pp. 356–72.

ARMSTRONG, J.M. (1985) 'A national assessment of participation and achievement of women in mathematics', in CHIPMAN, S.F., BRUSH, L.R. and WILSON, D.M. (Eds) *Women and Mathematics: Balancing the Equation*, Hillsdale, NJ, Lawrence Erlbaum Associates, pp. 59–94.

BELENKY, M.F., CLINCHY, B.McV., GOLDBERGER, N.R. and TARULE, J.M. (1986) *Women's Ways of Knowing: The Development of Self, Voice, and Mind*, New York, Basic Books.

BELL, E.T. (1937) *Men of Mathematics*, New York, Simon and Schuster.

BURTON, L. (Ed) (1986) *Girls into Maths Can Go*, London, Holt Educational.

COUNTRYMAN, J. (1992) 'Is gender an issue in math class? Perhaps it's time to change the subject', in THE NATIONAL COALITION OF GIRLS' SCHOOLS (Eds) *Math and Science for Girls*, pp. 72–85.

DAMARIN, S. (1994) 'Genders, mathematics and feminisms', Paper presented at the Annual Meeting of the American Educational Research Association, New Orleans, April 1994.

EQUALS (1989) *Assessment Alternatives in Mathematics*, Lawrence Hall of Science, University of California at Berkeley.

FENNEMA, E. (1994) 'Old and new perspectives: On gender and mathematics', Paper presented at the Annual Meeting of the American Educational Research Association, New Orleans, April 1994.

GILLIGAN, C. (1982) *In a Different Voice*, Cambridge, MA, Harvard University Press.

HANNA, G. (1989) 'Mathematics achievement of girls and boys in grade eight: Results from twenty countries', *Educational Studies*, 20, pp. 225–32.

HENDERSON, D. (1981) 'Three papers', *For the Learning of Mathematics*, 1, 3, pp. 12–15.

KAISER-MESSMER, G. (1994) 'Analysen "Frauen und Mathematik" — Nachbetrachtungen', *Zentralblatt für Didaktik der Mathematik*, 26, pp. 63–5.

KASEBERG, A., KREINBERG, N. and DOWNIE, D. (1980) *Use EQUALS to Promote the Participation of Women in Mathematics*, Berkeley, CA, University of California.

MCINTOSH, P. (1983) *Phase Theory of Curriculum Reform*, Wellesley, MA, Center for Research on Women.

MCINTOSH, P. (1989) *Interactive Phases of Curricular and Personal Re-vision With Regard to Race*, Wellesley, MA, Center for Research on Women.

OSEN, T.M. (1974) *Women in Mathematics*, Cambridge, MA, MIT Press.

ROGERS, P. (1985) 'Overcoming another barrier: Real women don't do math — With good reason!', *Canadian Woman Studies Journal*, 6, 4 (Winter), pp. 82–4.

TOBIAS, S. (1978, 1994) *Overcoming Math Anxiety*, New York, Norton.

Part 1

Effecting Change

In the introduction we provided an account of the approaches that were used in the first three phases of curriculum reform, as described by the McIntosh model, to increase the participation of women in mathematics. This part of the book focuses on transitional approaches (from Phase Three — Women as a problem in mathematics — to Phase Four — Women as central to mathematics). In their attempt to change the attitudes and behaviours of women towards mathematics we see many features typical of Phase Three approaches. However, the approaches included here, to different degrees, are distinguished by the extent to which they question monocultural assumptions and/or are influenced by some form of feminism. None of these approaches is firmly rooted in a Phase Three deficit philosophy that identifies gender with women and sees women as victims in mathematics education with problems for which they need help. Rather, these approaches seek to redress the gender imbalance in mathematics education by locating its causes outside women. The chapters of this part are organized according to their primary focus: female students, future teachers, practising teachers and the public.

Two mathematics camps described in this section intervene exclusively with female students and have as a common starting point the goal of improving girls' chances to succeed in mathematics by teaching them skills of survival. 'SummerMath', the summer learning community for secondary school girls described by Charlene and James Morrow, is designed to address the lack of challenge and support faced by many female students in learning mathematics. Drawing on constructivist learning principles and feminist scholarship on connected teaching, the programme provides a variety of experiences that connect girls to mathematics rather than alienating them from it. Denisse Thompson's chapter describes the 'Metro Achievement Program' which, like SummerMath, began as a summer option and has since expanded into a year-round multi-faceted intervention programme designed to enable disadvantaged girls of average ability reach their full potential, both academically and personally. Working outside the school environment, the programme augments the regular school programme by providing girls with more challenging and relevant mathematics experiences and opportunities to take risks and assume leadership roles.

Olive Fullerton and Lynn Friedman both focus on working with future teachers but their approaches derive from very different feminist perspectives and are situated in different phases of the McIntosh model. Fullerton's chapter details her work with a small group of prospective elementary teachers. In joint discussion, they reflected on their experiences in learning mathematics and the anxiety they shared. These reflections revealed, among others, that lack of experience and facility in using mathematical language and technical terminology were the main reasons these future teachers felt intimidated during mathematics lessons and as a result lost confidence in their mathematical abilities. Fullerton makes a series of recommendations, characteristic of feminism of difference (see Mura, Chapter 19), for changing the teaching process to benefit women's learning styles. This places her approach in Phase Four in contrast to that of Lynn Friedman's study which is in the transitional phase described in the introduction. Friedman's work springs from a review of recent studies and accounts which document that the proportions of women receiving graduate degrees in mathematics, and hence the proportion of women teaching in post-secondary institutions, are still too small. As conditions which foster success for women in graduate programmes, Friedman cites providing sensitive counselling, ensuring a critical mass of women students in graduate schools of mathematics, and having sympathetic and concerned women on the faculty. In contrast with Fullerton's approach, Friedman is influenced by liberal feminism in that she believes that by changing the conditions of women in graduate programmes (though notably *not* the women), the problem of low participation will be solved.

Three chapters place work with practising teachers in the foreground. Although interventionist by design, none of these approaches sees women as a problem in mathematics education. Rather they question the whole teaching and learning process. The first two papers describe the establishment of networks of educators and other adults who play a key role in the lives of female students. The Swedish network described by Barbro Grevholm uses the medium of conferences and newsletters to share information such as current research and teaching ideas, and to support and encourage members to work on changing the gender balance in mathematics education. The Québec 'Women Do Math' network developed by Claudie Solar, Louise Lafortune and Hélène Kayler, has similar aims, but in addition promotes teaching strategies and activities based on principles of feminist pedagogy. The focus of Cornelia Niederdrenk-Felgner's chapter is different. In working with in-service teachers she aims to increase their awareness of gender issues in the use of technology and the need to change the whole education system.

An entirely different approach is taken by Mary Harris in the final chapter in this section of the book. Her chapter describes the development of an exhibition, entitled 'Common Threads', which features the mathematics involved in women's traditional work. In its questioning of the conventional undervaluing of women's labour, usually labelled not-mathematical, and in her focus on uncovering the 'real' mathematics of the workplace, Harris's work is illustrative of Phase Four reform.

Chapter 2

Connecting Women with Mathematics

Charlene Morrow and James Morrow

A Summer Journey

Imagine yourself, a 15-year-old in the process of completing your second year in high-school, and one of six young women in a class of sixteen advanced-level mathematics students. One day, a brochure for something called 'SummerMath' arrives in the mail. Looking through it, first at the pictures, you find these images:

- three young women measuring a slab of rock;
- two students discussing what appears on a computer screen;
- two students (one with her shoes off) studying next to a blackboard;
- two students putting a robot together;
- a basketball court with a hard-driving woman dribbling toward the hoop;
- women together on a mountain top;
- four young women in a canoe; and
- two students intently at work in a laboratory.

Jumping out at you from the brochure appear these words from a former student: 'I learned to think "why" and "how" and not just to do as I've always been told. I only wish I had asked those questions of my teachers in high-school — but I will when I enter Mount Holyoke'. Juxtapose these words with what you heard from your teacher in school today, 'Just do it!'. Also staring you in the face are the words, 'an opportunity for young women'.

Well, that brochure came at the right time, and you sent off the application, sweating through the writing of an essay to gain admission. 'Good practice for applying to college', your parents said. Over the next several weeks, forms and information arrive from SummerMath preparing you for the summer away from home — for a 'college-like' experience.

The young woman beginning the journey described above is just one of over 1000 young women we have worked intensively with during the twelve-year history of the SummerMath programme at Mount Holyoke College. After a summer of constructing solutions to mathematical problems and learning to explain their problem-solving processes, many of these young women have taken charge of their mathematics learning in a variety of effective ways. Below we describe the structure and pedagogy of SummerMath and discuss the deep connections between our approach and new scholarship on women's education (Belenky *et al.*, 1986).

We begin our discussion with a brief summary of the state of mathematics education for women in the US. It is important to note that the body of literature from which the following information is drawn has focused largely on majority women. Clewell and Anderson (1991) have found that 'studies of women overlook women of color and that studies of students of color de-emphasize gender differences' (p. 3). While there are many parallels between the experiences of women of colour and white women, the nature of the similarities and differences has not been carefully studied. This issue is central to our concerns because the SummerMath programme serves students from many racial backgrounds.

Why 'An Opportunity for Young Women'?

Two decades of research on gender issues in mathematics education in the US have made it clear that there are still firm barriers preventing women's equal participation in mathematical studies and careers (Fennema and Leder, 1990; American Association of University Women, 1992; Sadker and Sadker, 1994). A recent publication, *Winning Women Into Mathematics* (Kenschaft, 1991), puts forth a detailed and convincing case that, as a culture, we still do a great deal to discourage women from participating in mathematics. Those individuals who have taken a careful look at the ways that girls experience our educational system tell us that girls are still receiving strong messages of exclusion, subordination, and objectification. Even in elementary school, where girls generally receive better grades than boys, we begin to see the effects of these barriers. Girls begin to perform more poorly on higher-level cognitive tasks in mathematics. Moreover, this difference seems to persist and grow through high-school (Hyde, Fennema and Lamon, 1990).

Across all grade levels, the attention afforded to girls in the classroom is far less than that given to boys (Leder, 1990; Sadker and Sadker, 1994). One particular aspect of this differential attention is that girls are not pushed to struggle for understanding. Usually, if a female cannot respond immediately to a question or problem, she is not offered guidance or encouragement by the teacher to persist in her thinking.

Women are confronted by many barriers that men do not have to face. For example, a disturbing number of students agree with the statement, 'Mathematics is more for boys than for girls'. Some studies found that such an attitude is pervasive among young males and that it holds, though to a lesser extent, among young females (Eccles and Yee, 1988; Morrow and Morrow, 1991). In addition, this research has also demonstrated that many high-school counsellors still discourage women from taking advanced mathematics courses and from considering technological careers, that teachers interact more with their male students than with their female ones, that boys, far more than girls, utilize computer resources during unstructured time, and that parents still consider mathematics, science, and technology to be more suitable subjects for males than for females.

In advanced high-school mathematics classes, girls begin to be in the minority (Czujko and Bernstein, 1989; Fennema, 1987). In college, despite the fact that women represent about half of the mathematics majors, they still find themselves in the minority in most of their mathematics classes due to the presence of a large number of men from male-dominated fields such as physics and engineering. In graduate school, the number of women declines drastically, to the point that men outnumber women four to one at the PhD level in mathematics (Billard, 1991).

Even for women who have chosen to dedicate themselves to mathematics, many barriers remain in place, especially at the most prestigious research universities where only 5 per cent of the entire mathematics faculty are female (Jackson, 1991). Indeed, successful educational environments for women are more often found in the liberal arts colleges than in the research universities. Liberal arts colleges, particularly women's colleges, contribute disproportionately to female PhDs in mathematics and engineering (Sharpe, 1992).

Career statistics for women indicate a vast disparity between the sexes in participation rates in quantitatively based fields (National Research Council, 1991). The more mathematics required in a particular career, the higher the pay and lower the rate of female involvement. The career picture for women in science and engineering fields is especially bleak. As of 1984, women, who represent 43 per cent of all professional workers, held only about 12.8 per cent of the jobs in science and engineering. Approximately 1 per cent of engineers, 2 per cent of physicists, and 5 per cent of chemists are women (National Research Council, 1991). The situation for women of colour is even worse (Clewell and Anderson, 1991).

Until very recently, female role models have consistently been written out of the history of mathematics (Perl, 1978; Perl and Manning, 1985). The one female, Emmy Noether, who is included in the 'Men of Mathematics' poster (produced by IBM and covering over 2000 years of history in mathematics) is devilishly hard to find. The culture of mathematics remains distant, cold, and undesirable for too many women.

The research literature is quite consistent regarding the factors associated with women's success in mathematics classrooms. Eccles (1987) describes 'girl-friendly' classrooms as having 'low levels of competition, high levels of cooperative learning or individualistic learning structure, high levels of teacher communication . . . of the intrinsic value of maths and the link between maths and various interesting occupations' (Eccles, 1987, p. 158). Pedersen (1988) adds that classrooms that are successful for females are 'not remedial . . . and employ a problem solving format in which students work on challenging problems' (Pedersen, 1988, pp. 209–10). Unfortunately, girls have rarely had the opportunity to reap the benefits of these findings. We, along with many others, conclude that a multitude of social factors discourage women from making mathematics a significant part of their lives.

SummerMath: Reform within a Feminist Framework

Let us now return to the summer journey with which we began our discussion, to the programme called SummerMath, held each summer on the campus of Mount Holyoke College. SummerMath is an intensive, six-week residential mathematics learning community for 100 high-school women. It is designed to address the ways in which women are underserved in mathematically based fields by providing new perspectives and new experiences of mathematics, computing, and science. The student body is very diverse, racially, geographically, and academically. In 1994, our student body was 37 per cent African or African American, 17 per cent Latina, 13 per cent Asian, 2 per cent Native American, and 30 per cent European American. They came from all across the mainland US, Hawai'i, Puerto Rico, Japan, Paraguay, Brazil, and Pakistan with ages ranging from 13 to 18 years. Some of our students have achieved good grades in mathematics, while others have not, but almost all feel the downward spiralling of self-esteem and confidence in mathematics that is so characteristic of high-school women (Eccles, 1987; American Association of University Women, 1992). We emphasize greater conceptual understanding, affirmation of young women as capable members of a learning community, and the importance of constructing one's own understanding of complex ideas.

The instructional methods, based on constructivist (von Glasersfeld, 1983) and connected models of learning (Belenky *et al.*, 1986), are designed to change the blind memorizing and rule-following behaviours that limit so many students' understanding of the underlying concepts of mathematics. SummerMath helps students replace such unproductive methods with more flexible problem-solving approaches. The atmosphere of the programme is one of challenge and support. That is, the challenge of rigorous study and difficult problems, coupled with the support of a community of teachers, residential staff, and peers. Our goal is to help students acquire greater confidence in their ability to excel, greater persistence in problem-solving environments, and a greater conceptual understanding of mathematics.

Staff Preparation

For one week before the students arrive, our staff of thirty-five meets to prepare. Our activities fall into three categories: doing mathematical and computer activities; training teachers, teaching assistants, and residence assistants; and consciousness-raising around issues such as racial and gender equity. It also includes practice and reflection on mathematical activities in order to build understanding and commitment to constructivist principles of active learning. Our focus is to look within ourselves for potential points of connection that can foster the growth and development of our students.

Classes

Students at SummerMath participate each day in classes characterized by student activity, questioning, discussion, and discovery. In all sessions, pupils begin by either solving problems, or posing their own problems. They experience a new vision of mathematics through two classes, 'Fundamental Mathematical Concepts' and 'Computer Programming' (LOGO), and three workshops.

'Fundamental Mathematical Concepts' is the heart of SummerMath. Students work and discuss ideas in pairs or small groups, with at least one instructor and one undergraduate assistant circulating about the classroom posing questions and facilitating discussion. The specially designed problem-based curriculum focuses on concepts essential to understanding advanced mathematics and science such as ratio, linear relationships, functions, patterns, and logic. Students are encouraged to assume an authoritative role as they construct their own understanding of mathematical concepts. Visual and verbal methods of illustration, as well as numerical, algebraic, and graphical representation, are stressed as means to deeper understanding. The frequent discussion among students, as well as between students and their instructors, provides a supportive environment that encourages reflection on thinking and the stimulus of collaborative effort.

In LOGO, students work in pairs at a computer and solve problems in the context of design projects. They pursue mathematical ideas ranging from elementary geometry to recursive functions. They learn to plan, organize, and revise their ideas by working on tasks such as transformational geometry, tangram puzzles, patchwork quilt designs, and group murals. In the last week of SummerMath, students demonstrate the programming skills they have developed by collaboratively designing, developing, and completing a major computer project.

Finally, each student takes three two-week workshops, choosing from a variety of options such as confidence building, brain imaging, the art of making anatomical comparisons, architecture, genetics, economics, psychology, and the physics of motion. These workshops provide experiences that place students in the role of scientists, thinking up the questions to be asked, performing experiments, collecting and analysing data, and comparing their observations with their hypotheses.

The Residential Community

The residential component of the programme is crucial in the creation of a strong, close-knit community at SummerMath. The residential staff is headed by a coordinator and an assistant, and includes several undergraduate residential assistants (RAs) who are also teaching assistants. The RAs give SummerMath students support and friendship as they face the challenges of learning and explaining their ideas in depth. In consultation with SummerMath participants, a series of workshops and social activities is developed, including issues of sexuality, nutrition and other health-related topics, dance and music, and t-shirt decoration and beadwork. Students are offered many additional activities intended to link their mathematics experience to the outside world, to give them a realistic view of career options, and to broaden their horizons. Once a week, speakers representing science, business, and various professions are invited to present an evening lecture, panel discussion, or laboratory demonstration and answer questions about their work.

Assessment

Assessment is embedded in the learning process through ongoing discussions with SummerMath staff. Students evaluate their own progress half way through the programme, and discuss their self-evaluation with their instructors. At the end of SummerMath, teachers comment on the progress of the pupils in their classes and make suggestions for further work.

Students write an extensive final evaluation of their summer experience, including feedback on all aspects of the programme. This evaluation is both an opportunity for each student to reflect on her own experiences, as well as a chance for us to learn how to improve the programme in the future. During the academic year following the programme, students complete a questionnaire investigating possible changes in attitude and new ways of approaching learning, as well as providing information about the courses they are taking, their grades, and standardized test scores.

Avenues of Connection for Women

Women's Ways of Knowing (Belenky *et al.*, 1986) gives a compelling description of the manner in which teaching can alienate women from most academic areas. Meaningful educational experiences described by women are those which permit them to weave together multiple aspects of their lives. The authors name this kind of educational experience 'connected learning', in that students are encouraged to build on their entire knowledge base rather than leaving all personal experiences at the classroom door. A teacher who facilitates this kind of education is engaging in connected teaching.

The following sections describe the journey from disconnection to connection in mathematics that we have observed many times in our students. Each section represents a critical issue for females and each is an integral component of our approach. These overarching issues are drawn largely from *Women's Ways of Knowing*. Within each segment, we describe the type of approach used in our programme to implement our educational goals.

Confirmation of Self in the Learning Community

Perhaps one of the most critical issues in education is that many women describe themselves as remaining on the fringes of the intellectual community. In mathematics, because women often feel completely on the outside, it is particularly important to structure the classroom as a setting with a place for everyone, where it is clear from the beginning that what students already know will be useful, that they will be treated as capable learners, and that there are no prerequisites for belonging.

At SummerMath, one way in which the self is confirmed is by having students write about and share problem-solving experiences and feelings (Buerk, 1985;

Morrow and Schifter, 1988). In this way, students come to see that there are many acceptable paths to knowing in mathematics. The teacher's role as facilitator and guide, rather than lecturer and expert, is critical to this process. The teacher must become skilled in active listening and asking questions that will allow the student to become more aware of her own thinking, as well as to decide which of her ideas to pursue further. This kind of teacher role has been likened to that of the Socratic teacher. However, a crucial difference for us is that the Socratic teacher maintains control of the dialogue mainly by asking leading questions and defining the terms of the conversation. At SummerMath, the student is in charge of the interaction and thus, the direction of talk is primarily defined by the pupil. The teacher, of course, inevitably provides a 'safety net', whereby misconceptions can be recognized and addressed. But, when the student is in control, rather than the teacher, it is much clearer to both parties when conceptual understanding has been attained.

Learning in the Believing Mode of Communication and Questioning

Many of the women interviewed by Belenky and her colleagues expressed a distaste for being in an argumentative atmosphere and, therefore, would often patiently await its end rather than participate. As a result, inquiry that is based on belief, rather than doubt, is essential to a classroom environment that serves women well. As our students share problem-solving experiences, it becomes the role of the teacher and other students to question why and to ask for details: Why does such a step make sense? What part of the problem allows for such a statement? What more can be said about the process used? Such questions are asked in the belief that the student has a valuable understanding that can be expanded, clarified, and modified.

Students are asked to provide detailed explanations for all solutions to problems through an inquiry process based on belief in the student rather than disbelief of what the student says. As each individual becomes used to explaining her thoughts and processes, she becomes more and more of a mathematician and a creator of mathematical knowledge. When a student can demonstrate to herself that she knows why a solution works, she becomes more confident and less teacher-dependent.

Taking on Challenges with Support

In order to grow intellectually, a student must take reasonable risks and be able to make mistakes. We, as a culture, have a difficult time allowing risk-taking behaviours in women. We 'rescue' females at the smallest sign of discomfort and help them into helplessness by providing the kind of support that takes control away from the individual. On the other hand, we give all students the message that you 'have to be tough', 'gut it out', or 'just do it' in order to get ahead. These messages convey challenge with no support and a sense of isolation, creating an atmosphere that many women avoid.

At SummerMath, our students work in pairs in all parts of the programme, including computer tasks. In the beginning, each student in the pair is given a specific role, either the problem-solver, or the active-listener/questioner in the believing mode described above. Students switch roles as they move from one problem to the next. In this way, they become better both at persisting in problem-solving, and in asking themselves questions that will help them to move along. Pupils also become better at asking specific and focused questions of teachers when help is needed. In order to become confident learners, students must struggle for understanding. However, they can struggle together and look to each other and the teacher for support and encouragement.

The Development of Voice

To help students gain a sense of their own voice in mathematics, we first help them move away from a strictly answer-oriented approach to solving problems. We have found that it is impossible to achieve this goal in an environment where answers are supplied. Therefore, we do not tell students whether they are right or wrong. Nor are students told how to solve a problem. When neither answers nor instructions are supplied, students are forced to look to themselves and to each other in order to devise a strategy for solving the problem. In explaining her solutions, the student must learn to give voice to her discoveries. We emphasize visual representations of solutions as an effective mediator between a student's ideas and her expression of them. Participants report that being able to 'see what I am doing' is a very powerful experience. The student can then give voice to what she sees.

Becoming a Constructor of Knowledge

As students work through mathematical challenges in a supportive learning community and develop a sense of voice and authority in the classroom, they begin to define their own questions. They become excited about the possibilities of posing their own problems and inventing new knowledge. They no longer remain outside, but become part of the inner circle of knowers, with their own power base. As this awakening begins, we try to provide insights into mathematical connections with other areas. Workshops in mathematically related fields such as architecture, chemistry, and medicine help students learn about the manner in which mathematics is applied in a variety of settings.

The Journey Home

> A few years after attending SummerMath, I enrolled in a college calculus course. In the final exam, I looked at the problems and quickly realized that I thought I could do all of the problems, but I would not be able to finish in the time allotted. I set to work, recalling the ways in which I learned to work confidently at SummerMath. When the time came to hand in papers, I, of course, was not done. I refused to give the professor

my paper and continued to work. I worked and worked until I had finished all of the problems, and felt confident that I had done them correctly. I handed my paper to my professor and said, 'I dare you to fail me'. I received an A! (Danielle)

We have heard many other stories of young women taking charge of their learning, from simply asking the kinds of questions that enable them to move ahead, to having the courage to take honours and advanced level courses. The data we have collected on attitude changes experienced by our students (Morrow, 1991) indicate that our pupils leave SummerMath feeling more able to persist in doing mathematics problems, more aware of the usefulness of mathematics, and dramatically more confident. Changes in student attitudes, beliefs, and behaviour are most effectively illustrated by students' own comments on their abilities to learn mathematics.

On beliefs about oneself:

- I think I have learned new techniques for solving problems, and have become more secure and confident. Now I know why certain theorems, etc. work.
- I learned you don't have to be a Mister Wizard to solve a confusingly worded problem.
- It makes me now, believe it or not, like problems that are hard. It also gave me patience and several ways to solve a problem. My thought processes will never be the same.
- I learned more about myself. I learned it's alright to speak up. It gave me courage and made me more relaxed in all sorts of situations.

On working with a partner:

- You get to see how other people think and how they're taught.
- You don't feel alone.
- Having to justify all of your answers.

On things that will be different in mathematics class:

- I'll feel I can depend on myself.
- The fact that I know there is more than one way to work a problem, if I don't understand the given way.
- I won't give up as easily when I get to a difficult problem; I will try different ways to solve problems instead of one way; I will strive more for good grades because I see how important mathematics is.

On experiences in fundamental mathematical concepts:

- I feel I'm capable of solving almost any math problem and if I stick in there I'll solve the problem, happily.
- I learned a lot about math, people, and the real world, good and bad.
- Self-satisfaction; better understanding; less frustrated when I don't under-

stand everything the first time.

- I learned a great deal about how I think about mathematics and solve problems, about the logic behind math procedures and how to rely on myself to solve problems. I also learned how to listen more carefully to my own ideas and how to find what my weaknesses are.

These comments illustrate that students take home with them new ways to approach mathematics that emphasize reasoning, comfort with the subject, motivation to study it further, a better understanding of certain fundamental concepts, and an overall increased self-confidence. They become independent, but not isolated, problem-solvers, retaining a sense of authority within themselves, but able to effectively communicate with others about mathematics.

An additional benefit participants gain is a much clearer sense of what a college environment is like and what kind of career choices may lie ahead. Our research on programme outcomes and resulting attitude changes (Morrow, 1991) indicates that SummerMath has proved extremely effective in furthering the mathematical and career aspirations of students from widely varying backgrounds, not just the mathematically precocious. Furthermore, analysis of data by racial category has shown that SummerMath has been effective for students of colour as well as white students, both in goals achieved during the summer and in positive attitude changes.

In such a strong, female-student community, intellectual and social components of friendships are strongly linked. Women don't have to live up to the expectations they believe males have of them. In SummerMath classrooms, there is an atmosphere of trust, cooperation, and support where women's voices are always heard, respected, and responded to. Teachers maintain high standards and push students to struggle for understanding in the belief that all pupils can use their personal experience effectively as well as justify their own ideas. In this atmosphere of belief, pupils become aware of, and trust, their own intellectual authority, and begin to construct their own knowledge and persist in problem-solving tasks. As an individual in the 1992 programme said,

> I learned so much about myself and how I interact with thoughts on a regular basis. I know I can be happy with myself and as long as I have that confidence in myself, I can take on challenges and not worry so much about failure. I will know that as long as I am thinking and working, I am succeeding.

The Case for a Single-Sex Mathematics Programme

Our belief in the virtue of a single-sex mathematics programme for girls is grounded in the long tradition of women's colleges, of which our institution, Mount Holyoke College, is a prime example. There is solid evidence (Riordan, 1990; Tidball, 1980) that women's educational institutions provide an atmosphere in which young women more successfully test their skills, find their voices, and go on to participate in a wide variety of arenas, including many that have been traditionally male-dominated.

It is interesting to note, however, that the mathematics education provided in the SummerMath learning community is not only in line with the structure of 'girl-friendly classrooms' described in the literature (Eccles, 1987), it is also consistent with that now recommended by the National Council of Teachers of Mathematics (NCTM, 1989) for *all* students in the US. The advent of new standards in secondary mathematics teaching, and the close attention being paid to the teaching of calculus at the college level (Steen, 1988), have brought about a great deal of positive change in mathematics education. There is now an emphasis on understanding rather than just memorizing, on communicating mathematics, and on doing mathematics in context. The publication of Professional Standards for Teaching Mathematics, by the NCTM in 1991, introduced the expectation that all secondary mathematics teachers, as well as teacher educators, will use much of the pedagogy we have been using at SummerMath. Why then, are we continuing to promote an all-female programme?

The fact is that pedagogical innovation in a coeducational environment does not automatically guarantee gender equity for the reasons discussed at the beginning of this chapter. This point is clearly illustrated by the following story of a middle-school teacher engaged in implementing some very positive changes in her classroom that are in line with the new NCTM teaching standards:

> Yet, even as this new atmosphere established itself in the classroom, Lisa became aware that some negative social norms continued to operate. Most dramatically, in this class of eight girls and fifteen boys, Lisa realized, several months into the year, that the girls in the class had no voice. Gender patterns that had been established in previous years of mathematics instruction, patterns reinforced by societal attitudes, continued to intrude ... Lisa had established a classroom dynamic of active and sometimes excited exchange of mathematical ideas, but the boys had come to dominate discussion. The girls were reticent about participating and thus were marginalized.
>
> Lisa saw a way to address the problem when, during a discussion with a mother who was concerned that her daughter had already given up on mathematics, Lisa suggested that her daughter and a friend stay after class on Wednesdays to do mathematics. But then, on second thought, she decided to encourage all eight girls to come, and they formed the 'girls' maths club'. The formation of the girls' maths club significantly altered the classroom dynamic. Lisa says, 'What the girls mastered ... changed ... their voice in the class. The voice of the classroom had been extremely masculine and what they did was to change the quality of that voice'. (Schifter and Fosnot, 1993, pp. 135–6)

Fortunately, this teacher could recognize and respond effectively to the traditional gender roles that remained in place despite her new pedagogical approaches. Given the strength with which gender roles resist change, we have little reason to believe that most educators will be able to see and intervene in this fashion without a great deal of encouragement and support. Teachers are rarely trained in explicit strategies for recognizing gender inequities or in the detailed approaches necessary to overcome them. It is one thing, for instance, to have an expectation for collaborative classrooms, but it is quite another to ensure that mixed-sex cooperative groups do not simply become another place where males do the talking and doing, while females do the listening and recording.

We do not wish to make the argument here that a single-sex environment is always best, only that it can and does serve young women well in some instances, particularly when gender messages are as strong as they have been regarding participation in mathematics and sciences. Neither do we wish to suggest that a single-sex environment is automatically an atmosphere in which young women can grow to their full potential. Historically, many women's institutions have been bastions for the conservation of women's traditional roles, functioning only to keep them in their place. To counteract unconscious conformity with societal expectations, there must be a purposeful effort to challenge gender stereotyping.

Strategies for Change

To make 'mathematics for all' a reality for women, we offer the following strategies based on our experiences with students and staff at SummerMath. While we believe that these approaches can often be implemented more effectively in a single-sex setting, we suggest that they are important considerations in any educational environment attempting to serve females well.

- Look at the experiences of female students on a regular basis. Being 'gender blind' is more likely to result in the continuation of inequitable patterns than in equal treatment.
- Look for the valuable aspects of female approaches to learning mathematics. It is too easy to fall into the trap of using successful male strategies as the norm for all.
- Develop a supportive professional climate for examining and challenging existing gender roles. Changing the classroom climate must be a collaborative effort by teachers and administrators. Educators in senior positions must be willing to assume a leadership role on this issue.
- Believe that female students thrive on intellectual challenge and that they deserve to be supported in their efforts.
- Become familiar with new scholarship on women, such as that done by Belenky *et al.* Cross-disciplinary thinking is rarely used to envision a more gender-inclusive educational environment.
- Challenge the personal beliefs of students, teachers, and administrators about the nature and usefulness of mathematics. Educating people about what mathematics is and does can create many new pathways for learning.

- Make sure that structures for small group work give students explicit strategies for sharing time and tasks equitably. It is likely that inequitable gender-based behaviour will continue and be reinforced in unstructured small group work.

Conclusion

It is a deeply ingrained idea, even within the broader feminist community, that mathematics is an inherently alienating experience for females. Many young women view learning mathematics solely as a stepping-stone to another place, a place where mathematics will seldom play a major role. We have too often heard the comment from young women that, 'I can do math, and it's even fun sometimes, but I want to choose a career that will allow me to do something useful with my life. I want to work with/for people.' Educators, parents, and mentors often impart the same view to young people, passing along the notion that, for instance, literature and social science are richly connected to human concerns and creativity, while mathematics is removed and mechanical. We believe that the SummerMath programme offers a model for helping young women see that they can do mathematics confidently, and that through mathematics they can become *connected to* people rather than *disconnected from* others.

References

AMERICAN ASSOCIATION OF UNIVERSITY WOMEN (1992) *How Schools Shortchange Girls*, Washington, DC, American Association of University Women Educational Foundation.

BELENKY, M., CLINCHY, B., GOLDBERGER, N. and TARULE, J. (1986) *Women's Ways of Knowing*, New York, Basic Books.

BILLARD, L. (1991) 'The past, present, and future of academic women in the mathematical sciences', *Notices of the American Mathematical Society*, 38, pp. 707–14.

BUERK, D. (1985) 'The voices of women making meaning in mathematics', *Journal of Education*, 167, pp. 59–70.

CLEWELL, B.C. and ANDERSON, B. (1991) *Women of Color in Mathematics, Science and Engineering: A Review of the Literature*, Washington, DC, Center For Women Policy Studies.

CZUJKO, R. and BERNSTEIN, D. (1989) *Who Takes Science? A Report on Student Course-Work in High School Science and Mathematics*, New York, American Institute of Physics.

ECCLES, J. (1987) 'Gender roles and women's achievement-related decisions', *Psychology of Women Quarterly*, 11, pp. 135–72.

ECCLES, J. and YEE, D. (1988) 'Parent perceptions and attributions for children's maths achievement', *Sex Roles: A Journal of Research*, 19, pp. 317–33.

FENNEMA, E. (1987) 'Sex related differences in education: Myths, realities, and interventions', in RICHARDSON-KOEHLER, V. (Ed) *The Educator's Handbook: A Research Perspective*, Longman Press, pp. 329–47.

Fennema, E. and Leder, G. (Eds) (1990) *Mathematics and Gender*, New York, Teachers College Press.

Hyde, J., Fennema, E. and Lamon, S. (1990) 'Gender differences in mathematics performance: A meta-analysis', *Psychological Bulletin*, 107, pp. 139–55.

Jackson, A. (1991) 'Top producers of women mathematics doctorates', *Notices of the American Mathematical Society*, 38, pp. 715–20.

Kenschaft, P. (Ed) (1991) *Winning Women into Mathematics*, Washington, DC, Mathematical Association of America.

Leder, G. (1990) 'Gender and classroom practice', in Burton, L. (Ed) *Gender and Mathematics: An International Perspective*, London, Cassell Educational Limited, pp. 9–19.

Linn, M.C. and Hyde, J.S. (1989) 'Gender, mathematics, and science', *Educational Researcher*, 18, pp. 17–27.

Morrow, C. (1991, April) 'A description of SummerMath, a process-oriented mathematics learning community, and resulting student attitudes toward mathematics', Paper presented to the National Council of Teachers of Mathematics, New Orleans, LA.

Morrow, C. and Morrow, J. (1991, October) *The SummerMath Program for Minority Girls and Teachers*, Annual Report, National Aeronautics and Space Administration Project.

Morrow, C. and Schifter, D. (1988) 'Promoting good mathematical thinking through writing in the classroom', *Proceedings of the Tenth Annual Psychology of Mathematics Education-North America*, DeKalb, IL.

National Council of Teachers of Mathematics (1989) *Curriculum and Evaluation Standards for School Mathematics*, Reston, VA, National Council of Teachers of Mathematics.

National Council of Teachers of Mathematics (1991) *Professional Standards for Teaching Mathematics*, Reston, VA, National Council of Teachers of Mathematics.

National Research Council (1991) *Women in Science and Engineering: Increasing their Numbers in the 1990s*, Washington, DC, National Academy Press.

Pedersen, K. (1988) 'Combating the stereotype of women in mathematics and women's stereotype of mathematics', *Proceedings of The Mathematicians and Education Reform Network, National Science Foundation Workshop*, Chicago, IL.

Perl, T. (1978) *Math Equals*, Menlo Park, Addison-Wesley.

Perl, T. and Manning, J. (1985) *Women, Numbers, and Dreams*, Santa Rosa, CA, National Women's History Project.

Riordan, C. (1990) *Girls and Boys in School: Together or Separate?*, New York, Teachers College Press.

Sadker, M. and Sadker, D. (1994) *Failing at Fairness: How America's Schools Cheat Girls*, New York, Charles Scribner's Sons.

Schifter, D. and Fosnot, C.T. (1993) *Reconstructing Mathematics Education: Stories of Teachers Meeting the Challenge of Reform*, New York, Teachers College Press.

Sharpe, N.R. (1992). 'Pipeline from small colleges', *Notices of the American Mathematical Society*, 39, p. 3.

Steen, L.A. (Ed) (1988) *Calculus for a New Century: A Pump, not a Filter*, MAA Notes No.8, Washington, DC, Mathematical Association of America.

Tidball, E. (1980) 'Women's colleges and women achievers revisited', *Signs: Journal of Women in Culture and Society*, 5, pp. 104–10.

von Glasersfeld, E. (1983) 'Learning as a constructivist activity', *Proceedings of the Fifth Annual Meeting of Psychology of Mathematics Education-North America*, Montréal, Canada.

Chapter 3

The METRO Achievement Program: Helping Inner-City Girls Excel

Denisse R. Thompson

Introduction

In the last decade, numerous reports have signalled the need to increase the participation in mathematics of students in North America (Mathematical Sciences Education Board, 1990; National Council of Teachers of Mathematics, 1989, 1991; National Research Council, 1989). These reports have indicated that, from high-school to graduate school, roughly half of all students at a particular level do not proceed to the next. For example, while approximately 50 per cent of all high-school students take more than two years of college-preparatory mathematics, only about 25 per cent take more than three years of mathematics at this level. For female students the problem is especially acute, since women in the US have traditionally taken fewer mathematics and science courses than males (Campbell, 1986; Meyer, 1989). However, by the year 2000 it is predicted that one-third of the total US workforce, as well as 85 per cent of new entrants to the workforce, will be women, minorities, and immigrants. If the workforce is to be a viable one, these groups need more and better mathematical training (Johnston and Packer, 1987).

Clearly, there is a need for broadly-based programmes that encourage all students to develop their abilities to their fullest potential. However, there are populations whose particular needs may best be served by programmes which focus specifically on their concerns. First, programmes serving female students should be developed which encourage them to keep their future options open by continuing with academic courses and this should begin early, for example, between 11 and 13 years of age. Nelson (1986) indicates that this is the period 'when girls are most vulnerable to peer values about what it means to be competent, successful, and female. Classroom and out-of-school experiences which consistently communicate stereotyped or limited expectations or are sparse in stimulating examples and encouragement will undercut girls' motivation to learn and will undercut the development of their full potential as productive and self-actualizing adults' (pp. 3–4). It is also at this age that girls' confidence in their own abilities and their expectations of success begin to decrease, factors which have been found (Campbell, 1986) to precede the decline in mathematics achievement by girls. These are the years when, according to Nicholson (1986), patterns of disengaging with mathematics and science become established. Thus, we need to design programmes to nurture female students. In single-sex environments female students are forced to assume leadership roles and develop cooperative networks. They cannot 'bow out' of difficult problems leaving them to the boys and so have to take the lead and solve problems for themselves.

The second population we need to target specifically are students of average ability. In the US, special education programmes are provided for students who have demonstrated strong academic promise and for those who are struggling and need remediation; few programmes exist to provide additional support for the large group of students in the middle. By encouraging those of average ability, it may be possible to foster links to strong academic programmes and keep such students in mathematics and science.

Third, in the US there is a need for programmes which serve students from inner-city neighbourhoods. These environments are home to a variety of racial and ethnic minority groups and much of the population is of a low socio-economic status. The schools often lack modern facilities and current materials (Matthews, 1984; McBay, 1990) and have high attrition rates (Arias, 1986). Therefore, support programmes are needed to enhance the educational experience of inner-city students and encourage them to remain in the system and to take advantage of academic opportunities.

The 'METRO Achievement Program'[1] was designed to address all three of these needs. Designed as an outside-school activity, the programme serves female students of average ability from the west side of Chicago. The programme aims to encourage academic achievement, especially in mathematics and science, to aid students' personal development, and to instill a spirit of community service. This chapter describes the programme and presents data on participants' perceptions of its benefits as well as their views on the programme being restricted to females.

Description of the Programme

The METRO Achievement Program began in 1985 as a summer programme for girls of average ability who were entering the seventh (age 12 years) or eighth (age 13 years) grades in the following fall. The programme was modelled on one for boys of the same age which had been operating in the area since 1962. Initially a tutoring and sports facility, it was expanded in 1972 to emphasize academic achievement and character development. When it was realized that girls needed similar opportunities, the founders made the decision not to make the existing programme coeducational since they felt that more could be accomplished with adolescents in a single-sex environment.

Although still focusing primarily on the middle-school grades 7 and 8, the METRO Achievement Program has expanded since 1985 to include activities throughout the academic year for girls in grades 4–12 (ages 9–17 years). At the present time, the programme consists of three segments. For girls between 9 and 11 years of age, there is an individual tutoring programme throughout the school year. Approximately 60 per cent of the tutors are professional women, 20 per cent are female college students, and the remaining 20 per cent are girls from the high-school METRO Program who devote one day per week to this activity. This provides the young girls not only with the tutoring they need but also with female role models. In the summer, girls in this age group may participate in a one-week mini-version of the programme offered to the older girls.

For 12- and 13-year-old girls, there is a five-week summer programme that concentrates on developing the whole person. Girls participate in a variety of academic courses, for example, mathematics with an emphasis on applications and technology, science centred on laboratory work, and communication skills including writing, grammar, and public-speaking. As well, girls are offered sessions on fine arts, team sports, personal counselling, and character development with a focus on interpersonal relationships. During the regular school year, participants complete a programme that is similar to that offered in the summer. Designed to upgrade skills, this programme aims to encourage students to enrol in an academic track rather than a general track with no college-preparatory courses.

At the high-school level, ages 14–17 years, the emphasis is on keeping girls in school and promoting enrolment in a college or university. The programme for the youngest girls concentrates on public-speaking as well as on the study of algebra emphasizing the use of graphing technology, graphics calculators and computer-graphing software. The curriculum for the following year focuses on writing skills using word-processing applications, and on the development of visual thinking skills using computer-generated geometric drawing tools. The 16-year-olds begin preparation for college-entrance exams, and in the final year the concentration is on the study of world culture. These curricular emphases closely follow the areas of study typically found in US high schools at this level.

The METRO Program recruits participants by making end-of-the-year visits to schools within its service area. After completing a survey indicating their interest in the programme, girls are chosen on the basis of follow-up interviews with applicants and their parents. During the selection interview, METRO tries to ascertain whether the girls' attitudes are in tune with the programme's goals. Staff also review their academic and attendance records, looking particularly at consistency of grades, comments on behaviour, lateness and absenteeism. Standardized test scores are also examined, with girls generally needing a score at least in the fortieth percentile. Essentially, the staff try to identify those good students who just need a chance. For a more detailed description of the selection process see Thompson and Jakucyn (1993).

During the summer, the programme serves about sixty 9–11-year-olds and eighty 12–13-year-olds. During the school year, there are approximately ninety girls aged 9–11 years, eighty aged 12–13 years, and eighty in the oldest age group (ages 14–17 years). The overall composition of the METRO Program intake is a critical feature. The diversity of the student population (60 per cent Hispanic, 30 per cent African-American, and 10 per cent Caucasian or Asian) gives girls the opportunity to interact with members of other racial or ethnic groups. For many students, this is the first time they have had the chance to make a friend with someone from another race.

Mathematics Component

The programme for the 12–14-year-old students uses materials developed by the University of Chicago School Mathematics Project (Usiskin, 1988, 1991). These materials were developed to provide students of average ability with a broader range of topics than is traditionally found in American textbooks. Topics added include statistics, probability, discrete mathematics, technology, and applications. Students begin with applied arithmetic and then move on to topics in algebra (such as patterning, variables, open sentences), pre-geometry (angles, polygons, area, perimeter), and more advanced problems in applied arithmetic.

There are several features that make these materials appropriate for use with the programme. First, students are expected to use scientific calculators, and these are made available to all participants in the programme. This ensures that lack of computational facility does not prevent the girls from studying mathematical topics they may not yet have met in their regular school curriculum. Further, the presence of technology allows for a focus on real-world problems with realistic numbers.

Second, 'real world' applications are integrated throughout the curriculum so that participants realize the importance of mathematics and its applicability to a wide range of situations. Applications are used to link branches of mathematics, such as algebra to geometry and statistics to algebra, and to bind mathematics to other disciplines such as social studies and science. Hence, an awareness of the need for mathematics is embedded in curricular emphases and students seldom ask 'When will I ever use this?'.

Third, students are expected to *read* the mathematics text. This is an especially important feature. Because METRO is a supplementary programme, it strives to help the participants develop the ability to become independent learners so that they can transfer learning from this programme back to the formal education environment. Thus, there are also assignments and periodical evaluations to ensure that students are learning and keeping pace with instruction.

The programme is staffed entirely by female teachers who have high expectations of participants and are convinced that all of the girls can and should learn a good deal of mathematics, for 'expectations of success often beget success' (N. Jakucyn, private communication). Teachers are expected to encourage an attitude towards mathematics that is contagious. The importance of attitudes in learning subject matter was addressed by Charles Merideth at a conference on mathematics for minorities:

> [O]ur success with subject matter is determined by our attitude and the

> attitude of those we are teaching. Attitudes have their own distinctive aromas; students can smell them ... They know when we care. They know when we believe in them, they know when we don't believe in them, and, in fact, they take their attitudes from our attitudes. (Merideth, 1990, p. 61)

Evaluation

From 1985 until 1988, only anecdotal evidence was obtained concerning the programme's benefits to participants. During the 1988 summer programme with the 12–13-years-old age group, a more formal evaluation was conducted. Girls and their parents were surveyed to determine the impact of the programme. Only a portion of the data is presented here; more detailed information about the evaluation can be found in Thompson (1989).

Thirty-eight parents completed the survey. When asked to comment on changes in their daughters that they would attribute to participation in METRO, about one-third of the parents indicated that their daughter was more determined and more sure of herself, and one-quarter indicated that she was better able to express herself. All parents indicated that they would recommend the programme to others.

Sixty girls responded to the survey. When asked about the influence of METRO, about one-fifth claimed that they had learned to have faith in themselves, 12 per cent commented that the programme helped them with their studies, and 10 per cent indicated that it was a good learning experience. The 13-year-olds felt that they were given insights on high-school and courses they should take. Participants were also asked to indicate whether the METRO Program had changed their attitudes towards mathematics. Almost half (45 per cent) of the respondents said that they had discovered that mathematics could be fun or that they could learn it. A tenth of the students claimed already to like mathematics when they joined the programme and reported no change in their attitudes.

During that summer, seven high-school students who were graduates of the summer programme were assisting as counsellors and teacher aides. Although a select group, these students were interviewed about their METRO experience and asked to reflect upon the impact the programme might have had on their lives. Six of the seven indicated that METRO had taught them they could achieve and be whatever they wanted to be. Four of the seven indicated that they had not originally planned to take mathematics and science throughout high-school, but that METRO had encouraged them to take more mathematics. Thus, for these girls the programme had the effect of influencing their academic preparation in high-school, and that preparation undoubtedly has increased the variety of options they will now be able to pursue.

In an effort to obtain more current data about the impact of the programme, a questionnaire surveying a wide range of issues was administered during the spring

Table 3.1: Influence of METRO on perceptions of mathematics (as a per cent of total responses)

	Middle-school		High-school	
	Agree	Disagree	Agree	Disagree
	(n = 36)		(n = 12)	
METRO has increased my interest in mathematics.	56	22	50	8
Much of the mathematics material at METRO has been new to me.	53	33	25	67
Before attending METRO, I was good at mathematics.	36	33	50	25
METRO helped me realize I could do algebra.	50	33	25	8
Because of METRO, I plan (am taking) college-prep mathematics courses in high-school.	33	17	33	8

of 1992 to a sample of the students participating in the academic-year session. Several of the respondents had been involved in some aspect of the summer programme, but others were new to METRO that academic year. The sample is small, so results should be interpreted with care.

Girls were given several statements about the influence of METRO on their perceptions of mathematics and were asked whether they agreed, disagreed, or had no opinion. Table 3.1 contains the percentage of responses to several of these statements for girls in the middle- and high-school years. For both groups, about half of the girls indicated that METRO had increased their interest in mathematics. Substantially more of the middle-school students than high-school students found the mathematics material new to them, perhaps a reflection that this is typically a level at which there is a great deal of review in the traditional US curriculum (Flanders, 1987). Furthermore, about half of the middle-school students indicated that METRO helped them realize they could do algebra, again, likely a reflection of the mathematics curriculum at METRO which places a great emphasis on pre-algebra, particularly patterns, use of variables, and solving equations. About a third of the participants either are taking or plan to take college-preparatory mathematics because of their METRO experience. These results are consistent with those obtained from the seven graduates of the programme who served as summer counsellors.

High-school participants were asked to comment on how METRO had contributed to the development of their mathematical ability. Table 3.2 contains a sample of responses. These girls saw METRO as encouraging them and helping them develop confidence in their ability to pursue mathematical tasks. When asked to comment on how METRO had hindered the development of their mathematical ability, there were no responses. However, one participant

Table 3.2: Responses to the question, 'How has METRO helped the development of your mathematical ability?'

- METRO helped me become more interested in math because of the way the[y] challenged me and showed me different ways of working out problems.
- I've always been good at math but METRO has pushed me even further to put more effort into what I was doing and help[ed] me get As instead of Bs.
- METRO has helped me in persevering and not giving up when it comes to hard mathematical questions.
- METRO has helped to strengthen my mathematical ability.
- METRO has taken the time to teach me math at my own pace.
- It [METRO] has given me easy ways to solve long word problems, it has also made me look at math in a positive way.
- They [METRO] helped me get interested in math.
- They [METRO] have helped me feel confident I could do the math work.
- METRO has helped me find strategies in taking math tests and has shown me some short-cuts to solving problems.
- [METRO helped me] find different ways of solving problems and it's more understandable.
- It hasn't helped much.
- I haven't done much math in METRO. I have however worked with the student I tutor.

Note: Comments are from high-school participants (ages 14–17) only.

mentioned that, although the junior METRO Program had enabled her to enrol in honours-level classes, she had been unable to remain in courses at that level.

As previously discussed, the creators of METRO had felt a need to design a programme solely to meet the needs of female students. In the teenage years, boys and girls begin to discover each other and it was felt that a coeducational environment would detract from the overall goals of the programme. However, no survey had ever been undertaken to this point to determine whether the participants thought the programme should include males. Table 3.3 contains the responses (agree, disagree, or no opinion) of twelve high-school participants to several statements about having male students included in the programme.

Most girls thought they would make as many friends, speak up as often in class, and receive as much individual attention if boys were a part of the programme. Nevertheless, half of the girls did not want to see boys included. Those who were against including boys thought they would not be as comfortable or as relaxed in learning if boys were present. Because the sample was small and consisted only of the high-school age group, care must be taken in extrapolating these comments to the larger METRO group or to other age groups. Still, the comments suggest that for some girls, single-sex programmes may be preferable. Such programmes may give some girls a chance to shine and take leadership roles, activities they would be less likely to engage in if the programme were coeducational.

Table 3.3: Reactions to including boys in METRO (as a per cent of total responses)

	Agree	Disagree
I would like to see boys allowed in the METRO program.	42	50
I would make just as many friends at METRO if boys were allowed in the program.	92	0
I would talk up in class just as much if boys were allowed in the METRO program.	67	25
I would get as much individual attention if boys were allowed in the METRO program.	58	17
I would learn as much in my METRO classes if boys were allowed in the program.	42	42
Boys are more disruptive in class than girls.	67	17
I would have liked METRO as much if boys were allowed in the program.	58	33
I would have liked METRO better if boys were allowed in the program.	8	42
Male teachers at METRO would care about me as much as female teachers.	42	33

Note: Percentages are based on a total of only twelve responses from high-school participants (ages 14–17).

Conclusion

The METRO Achievement Program provides a model for programmes designed to encourage young females to develop their full academic potential. Too often, students of average ability, especially if they are also female, are lost in the shuffle in US secondary schools and are permitted to drift through school without taking any academically challenging courses. The decisions girls make between the ages of 11 and 17 years can have far-reaching consequences for their future educational and career opportunities. The problems are especially acute for minorities and those from the inner-cities where drop-out rates are high. At a 1990 Education Summit, the following goal for US education was outlined:

> What our best students can achieve now, our average students must be able to achieve by the turn of the century. We must work to ensure that a significant number of students from all races, ethnic groups, and income levels are among our top performers. (US Department of Education, 1990, p. 3)

The METRO Program, which combines academic work with personal attention and encouragement, provides one avenue to help students meet such a goal and accept academic challenges.

While additional long-term evaluation is needed, that which has been conducted to date, both formal and informal, suggests METRO has a positive influence. High expectations by the staff seem to help participants develop confidence in their ability to succeed. Although all girls may not need such single-sex programmes, a substantial number of girls seem to need and thrive in such environments. In striving for equity in mathematics education, it is important that we meet these students' needs and provide an environment in which they can develop their full potential.

Note

1 I wish to acknowledge the helpful comments provided by Natalie Jakucyn of the METRO Achievement Program during the preparation of this chapter.

References

ARIAS, M.B. (1986) 'The context of education for Hispanic students: An overview', *American Journal of Education*, 95, pp. 26–57.

CAMPBELL, P.B. (1986) 'What's a nice girl like you doing in a math class?' *Phi Delta Kappan*, 67, pp. 516–20.

FLANDERS, J.R. (1987) 'How much of the content in mathematics textbooks is new?', *Arithmetic Teacher*, 35, pp. 18–23.

JOHNSTON, W.B. and PACKER, A.E. (Eds) (1987) *Workforce 2000: Work and Workers for the Twenty-first Century*, Indianapolis, IN, Hudson Institute.

MATHEMATICAL SCIENCES EDUCATION BOARD (1990) *Reshaping School Mathematics: A Philosophy and Framework for Curriculum*, Washington, DC, National Academy Press.

MATTHEWS, W. (1984) 'Influences on the learning and participation of minorities in mathematics', *Journal for Research in Mathematics Education*, 15, pp. 84–95.

MCBAY, S.M. (1990) 'Education that works for minorities: An action plan for the education of minorities', in MATHEMATICAL SCIENCES EDUCATION BOARD, *Making Mathematics Work for Minorities: A Compendium of Program Proceedings, Professional Papers, and Action Plans*, Washington, DC, Mathematical Sciences Education Board, pp. 30–3.

MERIDETH, C.W. (1990) 'Banquet Address, in Mathematical Sciences Education Board', *Making Mathematics Work for Minorities: A Compendium of Program Proceedings, Professional Papers, and Action Plans*, Washington, DC, Mathematical Sciences Education Board, pp. 56–65.

MEYER, M.R. (1989) 'Gender differences in mathematics', in LINDQUIST, M.M. (Ed) *Results from the Fourth Mathematics Assessment of the National Assessment of Educational Progress*, Reston, VA, National Council of Teachers of Mathematics, pp. 149–59.

NATIONAL COUNCIL OF TEACHERS OF MATHEMATICS (1989) *Curriculum and Evaluation*

Standards for School Mathematics, Reston, VA, National Council of Teachers of Mathematics.

NATIONAL COUNCIL OF TEACHERS OF MATHEMATICS (1991) *Professional Standards for Teaching Mathematics*, Reston, VA, National Council of Teachers of Mathematics.

NATIONAL RESEARCH COUNCIL (1989) *Everybody Counts: A Report to the Nation on the Future of Mathematics Education*, Washington, DC, National Academy Press.

NELSON, B.S. (1986) 'Welcome: Perspective on the Issues', *Proceedings from the Operation SMART Research Conference*, Indianapolis, Girls Clubs of America, Inc., pp. 3–5.

NICHOLSON, H.J. (1986) 'But Not Too Much Like School: Case Studies of Relationships between Girls Clubs and Schools in Providing Math and Science Education to Girls Aged Nine through Fourteen', Paper prepared for the Operation SMART Research Conference, Girls Clubs of America, Inc.

THOMPSON, D.R. (1989) *METRO Achievement Program: Summer 1988 External Evaluation Report*, ERIC Document Reproduction Service ED 317 651.

THOMPSON, D.R. and JAKUCYN, N. (1993) 'Helping inner-city girls succeed: The METRO Achievement Program', in CUEVAS, G. and DRISCOLL, M. (Eds) *Reaching All Students with Mathematics*, Reston, VA, National Council of Teachers of Mathematics.

US DEPARTMENT OF EDUCATION (1990) *National Goals for Education*, Washington, DC, US Department of Education.

USISKIN, Z. (1988) 'The sequencing of applications and modelling in the University of Chicago School Mathematics Project (UCSMP) 7-12 Curriculum', in BLUM, W., BERRY, J., BIEHLER, R., HUNTLEY, I., KAISER-MESSMER, G. and PROFKE, L. (Eds) *Applications and Modelling in Learning and Teaching Mathematics*, Chichester, Ellis Horwood, pp. 176–81.

USISKIN, Z. (1991) 'Building mathematics curricula with applications and modelling', in NISS, M., BLUM, W. and HUNTLEY, I. (Eds) *Teaching of Mathematical Modelling and Applications*, Chichester, Ellis Horwood, pp. 30–45.

Chapter 4

Who Wants to Feel Stupid *All* of the Time?

Olive Fullerton

Recently I found myself waiting in a long queue in a university library. When I finally stood opposite the student who was performing check-out service, the young woman stopped and stared for several moments at the title of the top book in the stack I handed her. That text happened to be Buxton's (1991) *Math Panic*. 'That was the best thing about getting out of high school', she said to me as she looked up. 'What was?' I asked. 'Never having to be involved with math again. Ever!' I asked about the source of her dislike for mathematics. 'Who wants to feel stupid *all* of the time?' she responded.

I teach a mathematics methodology course for university students preparing to teach elementary school children. In her distaste for the discipline of mathematics, the young woman performing library check-out duty exemplifies many of the student teachers with whom I come in contact. In fact, over the past ten years, my unofficial records of the attitudes towards mathematics of these emerging teachers show that, in every year at least 60 per cent, and sometimes as many as 70 per cent, of them express some degree of discomfort when it comes to teaching mathematics. Because of the very high proportion of females in my classes (in my most recent class all of the students were female), this means that, as Countryman (1992) has also observed, gender remains a thorny issue in the mathematics classroom.

One of the most troubling features of dysfunctional attitudes toward mathematics is loss of self-esteem. Embedded in the comment, 'Who wants to feel stupid *all* of the time?', are feelings of helplessness and despair. While it must be acknowledged that some gains have been made in helping women feel more confident and comfortable in the field of mathematics, we are still a long way from removing the well-established classroom barriers and practices, as well as the societal biases, which make it impossible for female students to function on an equal footing with their male colleagues.

Since the phenomenon of low confidence continued to be evident year after year among the female teacher-candidate population, I decided to carry out a study aimed at exploring how such low self-esteem evolves in this otherwise confident and competent group of young women. I wanted to discover whether it would be possible to alter the attitudes of individual students, and whether a change in teaching practices would enhance their mathematics-related self-confidence. Over a period of two months, I met regularly with six young women who volunteered to be involved in my study and who identified themselves as having low self-concept in their ability to 'do' and teach mathematics.

Our first meeting took an unexpected turn when the participants expressed extreme concern over what would be done with the information I would glean from the meetings. In the discussion which followed, their fears became very obvious, culminating in their refusal to allow me to tape or even to make notes of our conversation during the meetings. Heather summed up the feelings of the group when she said, 'No school board would hire me if it were known that I can't do fractions at all or that I don't understand any geometry. What parents would allow their children to be in my classroom? They'd have to be crazy. Their kids would turn out just like me and that would be *bad*.' I was however given complete liberty to make notes after our meetings and to use the written reflections which they gave me at the conclusion of each get-together. Apparently, they felt in control of what they recorded themselves. These reflections and my own notes formed the data for my study.

At this first meeting it also became clear that, for any confidence-building to be possible, we would have to narrow our goals and focus on a small area of mathematics. After much deliberation, there was general agreement that 3-dimensional geometry was a particularly troubling area for each student and so this was the strand of mathematics on which we decided to concentrate. The participants spent the final hour of the session writing. They recorded their feelings related to mathematics in general and wrote in particular about any incidents they could recall which they believed may have affected their attitudes toward mathematics. Over and over again their statements told how their self-concept had been diminished by thoughtless comments or inappropriate behaviours. For example, Julie wrote:

> I feel my negative feelings towards mathematics began when I was in the fourth grade [9 years of age]. I remember I was learning fractions and I was finding the exercises in the text book very difficult. Many times I stayed after school and I received extra help from my teacher. I think it was around the third or fourth day of the extra help when my teacher said to me, 'You're hopeless. You'll never understand this. It's just a waste of my time.' From that time on, I have always considered myself to be stupid in mathematics. . . . I expected it [failure] to happen and I wasn't surprised or shocked. I learned to accept my inability to understand mathematics.

The other participants had similar stories to tell. Their comments indicated a profound lack of self-confidence and poor self-esteem related to understanding, and the 'doing' and teaching of, mathematics.

As a group, we agreed on a definite procedure for each meeting. Each week, I would distribute an article which I felt was relevant to the area of geometry identified. The participants would read the article and respond to it in writing recording their feelings, any 'new' mathematics they learned as they read, any memories the reading stimulated, questions which came to mind, and so on. In our meetings, we would discuss any feelings induced by the article, and clarify any problems which emerged from reading their written comments. At the conclusion of each meeting, a few minutes would be devoted to having participants record any observations they wanted to make.

Over the course of our meetings, four issues pertaining in particular to gender and mathematics came to light. The first issue was that the participants didn't know how to 'talk math'; not one of them had ever been engaged in any form of mathematical discussion. Males had dominated the classroom mathematics discourse. The second issue was that they did not know the appropriate mathematics 'register' (Halliday, 1978) needed to carry on a normal mathematics discussion. Third, beyond basic arithmetic skills not one of these students could see any relevance for exploring the discipline of mathematics. Last, the issue of confidence was one which permeated all of our work; all of these young women felt 'stupid', had feelings of incompetence reinforced by teachers, male siblings, parents and members of their communities, and felt certain that they were beyond help as far as learning mathematics was concerned. While these issues are definitely connected, for purposes of clarity I will separate them and make recommendations pertaining to each.

Talking Mathematics

Our second meeting was a great revelation for me. The students had faithfully read the assigned article and made extensive notes which they guarded rather secretively. When it became apparent that their jottings were all similar in nature, they gradually relaxed and a much more open atmosphere prevailed in subsequent gatherings. At first we had difficulty establishing a natural flow of conversation. When I asked them to comment on the article, there was silence. It soon became evident that these young women were afraid of 'talking mathematics'. The fact that I was asking them to express ideas, opinions, concerns and problems was totally foreign to them. This, then, was our first hurdle; to become comfortable with the idea of talking in a situation where mathematics was being discussed. Before discussing the article, they found it necessary first to discuss their experiences in mathematics classes. Talk was unknown to them in that setting. Julie reported:

> I've never talked about mathematics before except to say that I can't do it. I've never actually talked in a mathematics class. I wouldn't know what to say. You know — I don't have anything mathematical to say.

After similar opinions had been expressed by several group members, Heather recalled that,

> When I was in elementary school I used to try to answer questions. I'd

> wave my hand and wave it but the teacher didn't notice me. Lots of girls had that happen to them. I guess maybe the teacher knew that we weren't any good. We used to talk about that. Once all of my friends [girls] decided we wouldn't put up our hands at all to see if the teacher would ask us questions then. She didn't even notice that we had stopped participating.

In her written response to the first article, Jill commented:

> I agree strongly with the author's suggestions that children should discuss and even argue about their findings as they explore shapes. I will always remember a day in grade 12 when I started crying in mathematics class because I couldn't understand what we were doing. A friend who had been helping me a lot throughout the year moved his desk next to mine, asked what I didn't understand and proceeded to explain the concepts to me. Usually when we talked our teacher just told us to be quiet but that day he sent both of us to the office for talking in class and the next day my friend had to sit in the seat farthest from mine. Reading the article brought this memory back to me and at the same time opened up a whole new way of thinking about talking and discussing! . . . Who knows? Maybe if I had been allowed to talk I would have walked out of math class that day understanding instead of spending my time in the Principal's office.

These student responses provide concrete illustrations of research findings with respect to the significance of language in the learning process. Take, for example, Vygotsky (1978):

> Our experiments demonstrate two important facts:
> (1) A child's speech is as important as the role of action in attaining the goal. Children not only speak about what they are doing; their speech and action are part of *one and the same psychological function*, directed toward the solution of the problem at hand.
> (2) The more complex the action demanded by the situation and the less direct the solution, the greater the importance played by speech in the operation as a whole. Sometimes speech becomes of such vital importance that, if not permitted to use it, young children cannot accomplish the given task.
> These observations lead me to the conclusion that *children solve practical tasks with the help of their speech, as well as their eyes and hands.* (Vygotsky, 1978, pp. 25–6)

The idea that language plays a significant role in learning is not a new one. Dewey (1916) wrote about the 'importance of language in gaining knowledge' and issued the caveat that 'education is not an affair of "telling" and being told but is an active and constructive process involving appropriate language with direct experience' (p. 17). More recently, Barnes (1971) indicated that the language of the classroom must be viewed as more than the medium of classroom communication; language must be seen to be part of the very learning process.

While there is extensive research and writing on the close connection between language and learning, the more specific question of the relationship between language use and learning mathematics has received less attention. There is, however, a growing body of research which makes evident that, to some extent, mathematics learning is also dependent on learners being given opportunities to 'talk mathematics'. Cockcroft (1982), in asserting that 'The ability to "say what you mean and mean what you say" should be one of the outcomes of good mathematics teaching', emphasizes the significance of group discussion during mathematics learning. Bishop (1985) contends that there is far too little opportunity for students to talk about and share mathematical ideas. Traditionally learners are silent in mathematics classrooms. Teachers talk and children listen or respond in one-word phrases. Why is this significant? Bishop (1985) believes that, 'A new idea will be meaningful for a pupil to the extent to which it connects well with the learner's existing ideas and meanings' (Ibid., p. 27). In other words, mathematics classes must become places where mathematics is discussed. Bishop urges teachers to make way for a dramatic increase in the amount of classroom mathematics 'talk'. Hoyles (1985), while stressing the importance of learner–learner discourse as opposed to teacher–learner discourse, also points out that becoming an active listener in mathematics is equally important and conjectures that, in the 'silences' of discussion, real learning is taking place, 'Listening for learning is not passive; it is an active attempt to incorporate another's schemes into one's own; it stimulates one to go outside oneself and look again at the arguments'. (Ibid., p. 207).

The message of these studies is consistent and clear. Ideal learning conditions for mathematics include opportunities to discuss the concepts under consideration. Yet not one of these young women recalls having 'talked mathematics' at any point in elementary or secondary school. The one student who did attempt to discuss a problem situation with her friend found herself humiliated by being sent to the Principal's office. Are the learning conditions experienced by these emerging teachers anomalous, or are they representative of learning conditions generally found in classrooms? In a Ministry of Education for Ontario (1990) review of grade 6 (11–12-years-old) achievement in arithmetic, geometry and problem-solving, the preponderance of children reported that they most frequently worked on mathematics alone and that their teachers most frequently worked or demonstrated on the blackboard. From this, one can infer that 'talking mathematics' has low priority in most Ontario grade 6 classrooms. Similarly, O'Brien and Richardson (1987) state that they found little acceptance, on the part of practising teachers, that language use is an integral part of learning mathematics. Desforges (1989) reports that, 'Fruitful discussions are rarely seen in primary classrooms. This is so even in the classrooms of teachers who recognize and endorse their value. When discussions are seen, they are mainly teacher dominated, brief and quickly routinized' (p. 150).

While the women in the research project didn't recall talking in their mathematics classes, they did remember that the males in their classrooms had many more opportunities to respond to mathematics questions posed by their teachers than they did. This pattern of classroom interaction in which males are preferred is well documented (Sadker, 1984; Coulter, 1993; Best, 1983). Fullerton (1993) reports that when a male in an elementary classroom was asked why he thought his teacher asked him more questions than the girls, he replied, 'She knows I know the answers and she knows that they don't.'

Mathematics Register

There are many situations in our lives from which special language patterns emerge. This expected and accepted manner of expressing ideas in a particular situation is often referred to as 'register' (Cazden, 1988; Halliday, 1978; Pimm, 1987). Halliday (1978), the first to use the word 'register' to refer to these specialized language patterns, characterizes the term in this way: 'A "register" is a set of meanings that is appropriate to a particular function of language, together with the words and structures which express those meanings' (p. 65). One well-studied and documented 'register' is 'baby talk' (Cazden, 1988), the special language used by mothers when they communicate with their infants and toddlers. As well, Cazden (1988) designates as a 'register' the specialized language of particular professions — the way doctors, lawyers, sports announcers, or auto mechanics talk when they are 'on the job'. Thus there is a medical register, a sports register, and so on. Access to appropriate registers and a degree of comfort in comprehending and using them can be extremely significant to those wishing to participate in, or understand, the relevant activities in particular situations. Mathematical language is also a 'register'. Pimm (1987) points out that 'It is not just the use of technical terms which sound like jargon to the non-speaker (of mathematics), but certain phrases and even certain modes of arguing that constitute register' (p. 76). Further, he insists that acquiring control over the mathematics register is a significant part of becoming a competent mathematician — at any level.

During our meetings, it became clear to me that one of the conditions preventing these six teacher candidates from discussing mathematics and from becoming 'competent mathematicians' was their lack of facility in using the essential mathematics 'register'. As reported earlier, when the idea of discussing mathematical concepts was introduced to the group, Jill indicated 'I wouldn't know what to say — you know, I don't have anything mathematical to say.' Heather even reported that 'When I try to say them, mathematical words just stick in my throat — you know — they just won't come out like regular words.' In other words, these young women were reporting that the language necessary for engaging in mathematical discussion had, to date, eluded them. In their written responses to the weekly readings, comments regarding the difficulty of the language or 'register' were common. Patricia's comments reflected an awareness of the need for a special mathematics register: 'The vocabulary [of the article] was slightly mind boggling; I compared it [the vocabulary] to the psychology vocabulary I'm learning and found that perhaps any discipline requires its own specific terminology and this wasn't a "bad thing" in itself.' Compare this statement to one she wrote several sessions earlier:

> Dodecahedron, icosahedron, ... what vocabulary! How did these labels develop? Why do they have to be so complicated sounding? I can never remember all these elaborate labels! ... Another point of dread arrived when this statement was made: '... simple quadrilaterals, for which no two sides intersect in any point other than an end point of both sides ...' What's so scary about that, you may well ask. I DON'T KNOW. I just hate the way it is all put together.... I really start to feel out of my territory.

When we talked about what, in fact, was 'so scary', all of the participants agreed that their fear was likely connected to not having actually voiced any mathematical words out loud. Julie reported:

> When I said 'the quadrilateral' today in our discussion, it was the first time I'd ever said it. Now that I've said it TWICE I think I can make it part of my everyday language. But last night when I was reading the article and came across the word for the first time in years, I found that I wanted to look away ... it was as if I didn't even want to see it, let alone say it ... it was almost as if it was obscene.

Making Mathematics Relevant

In recent years there has been a renewed emphasis on presenting the essential information of various disciplines through exposure to 'real' situations. In mathematics, more often than not, this has been accomplished through such gimmicks as inserting children's names in otherwise artificial problems or by having children create their own problems. The purpose of making these so-called connections is to enable students to 'see' the significance of their hard work and, as a result, achieve greater satisfaction from knowing that they can employ those newly acquired skills in 'real world' situations. My criticism of this approach should not be taken as suggesting that schools should teach only mathematics that is needed in life, for who can predict with any certainty what might be useful in the future. Even if crystal-balling were reliable, I would not advocate that mathematics be reduced to a utilitarian discipline. Rather, I am suggesting that it might be possible and desirable for the mathematics of elementary and high-school to provide starting points in life situations familiar to students, and for learners to be assisted in fitting new mathematics into already established patterns of thought — in learning to make connections with what is already known.

One of the barriers to understanding mentioned frequently by the teacher candidates working with me, was that, for them, mathematics was neither relevant nor meaningful. They did not appreciate that mathematics permeated their lives, that their every action was in some way connected to mathematics, that the beauty and harmony of their world was due in large measure to mathematics. Countryman (1992) writes that, 'More often than not they [women] see a mathematics that contains neither women nor the concerns of women' (p. 75). Laura wrote passionately about the fact that mathematics did not 'connect' to any part of her life nor to any knowledge already within her grasp.

> I needed to know why . . . we were learning mathematics . . . More often than not I found math to be illogical and useless; what was the point of learning about a right-angled triangle? Was I going to bump into one while walking my dog down the street?

Once the participants realized that they could 'read' mathematics and began to feel slightly more comfortable with geometric concepts, their attitudes changed and they wrote positively about their new understandings. Patricia's comments are typical:

> Overall, I enjoyed discovering that math . . . can be applied to everyday life. I especially enjoyed seeing the picture of the child involved with the model of the bridge . . . I had never known about the rigidity of the triangle, nor the strength of the cylinder . . . This is the way math should be presented.

In discussion, Doreen added, 'At least now I can connect this mathematics [elementary geometry] with things I know already and with my way of learning things. I don't need to be "drilled" on this interesting stuff. I seem to understand it and remember it naturally.' In a recent article, Joan Countryman (1992) asks several important questions related to relevance:

> What would high school mathematics be if women were central to our teaching and thought? How would the metaphors change? Would we still speak of 'mastery', 'drill', and 'attacking problems'? Would the emphasis still be logic, abstraction, rationality, and right answers, or would it become construction, exploration, intuition? Would it shift from product to process? From exclusion to inclusion? (Countryman, 1992, p. 73)

These young women seem to have provided partial answers to at least some of these significant questions. Clearly, they felt the need to learn in more 'natural', exploratory, constructivist ways.

The Confidence Factor

There was one recurring theme in all our discussions and in every written response; the teacher candidates spoke and wrote about how, in relation to mathematics, their self-worth had become diminished over the years, how particular teachers had shattered their self-confidence, and how lack of success had destroyed their self-esteem. Their comments were filled with negative predictions about the success they expected to achieve in the group and their ability to help children be successful in learning mathematics. They spoke openly about the fact that they did not ever expect to be other than incompetent even with the most elementary mathematics. Doreen seemed to speak for the group when she said, 'I'm only telling you this at the beginning so that you won't get your hopes up that I'll learn some math and then be disappointed when nothing changes for me. Likely these others can learn but not me.'

Following the reading of the first article, we talked about the concepts presented for more than two hours during which I gradually introduced the mathematics register that was essential for the discussion. Not once did the intensity of the discussion fade or the focus shift. As well, to reinforce the ideas, we spent time examining appropriate concrete materials — models of simple geometric solids. Each participant agreed to ask questions 'no matter how dumb'. I tried to be encouraging and gave each student every opportunity to respond to one another's questions whenever possible. At the conclusion of that session each participant had, what she believed to be, a thorough understanding of the elementary concepts introduced in the article and a grasp of the mathematics register essential for discussion of those concepts. Each participant had selected a solid, described it using appropriate mathematics register, suggested where she had encountered models of the solid in the 'real' world and indicated how her concerns had been alleviated and in what ways she still felt apprehensive about this particular area of mathematics. Following this particular session, Jill wrote, 'It is a small step but it feels like quite an accomplishment . . . I feel that my knowledge is expanding, slowly but surely I feel that the very small mental framework I have for geometry is growing to accommodate a whole new wealth of information.' In a similar vein, Doreen indicated that 'I know this increase in knowledge will positively affect my confidence when teaching geometry to my class.' In the oral discussion, there was even a change of tone in their voices. They sounded more confident, more optimistic. 'Maybe I can learn this stuff after all', Julie commented.

Recommendations

What can we learn from this study about how the teaching and learning of mathematics affects female students? I believe it shows that we have to work on a number of areas concurrently. First, we must make teachers more aware of the significance of language in the learning process and particularly of the importance of language in the learning of mathematics. Mathematics learning experiences should become 'more like messy conversations than like synoptic logical presentations of conclusions' (Lampert, 1992, p. 307). Further, as Desforges (1989) has suggested 'If discussions are to be standard practice, then quite clearly the time will have to be made to conduct them in an undistracted fashion. This entails reducing drastically the breadth of the curriculum that teachers feel obliged to rush over' (p. 150). As well, teachers must be helped to realize that all learners in the classroom, not just males, must be provided opportunities for participating in mathematics discourse.

Closely connected to mathematics discussion is the issue of mathematics register. Learners of both genders must be given opportunities to use the register of the mathematics being discussed. Since learners often require demonstration, guidance and direction from 'experts' before they can function independently, teachers must 'talk mathematics' so that learners can do the same. Once they are familiar with the 'register', students must then be provided with many occasions which require them to use the appropriate mathematical language. But let me emphasize again, females must have equal access to the concepts of mathematics and the power and understanding that talking about them brings.

The third message extracted from the weekly study sessions is that relevance is of particular importance to girls and young women. Several years ago, I interviewed sixty children as part of a research project which attempted to assess the 'condition' of mathematics education in Ontario schools. When I asked where the students would use the mathematics just taught by their teachers, the following was a typical response: 'Oh, you wouldn't use that out there Miss [in the community], you only use that in here [the classroom].' The young women who met with me must have had similar classroom experiences, for not one of them recognized the relevance of mathematics in their lives. How can mathematics be made more relevant? Resnick (1987) discusses and supports the view that the focus of school mathematics must move away from 'symbols correctly manipulated' and 'decontextualized skills . . . divorced from experience'. Schoenfeld (1988) reminds us that:

> Whether or not it is intended, the students' sense of 'what mathematics is really all about' is shaped by the culture (context) of school mathematics — the environment in which they learn these facts and procedures. In turn, the sense of what mathematics is really all about determines how (if at all) the students use the mathematics they have learned. . . . [M]oreover, . . . the facts and procedures students learn in mathematical instruction should be a means to an end rather than an end in themselves. (Schoenfeld, 1988, p. 82)

If we follow this reasoning, we have serious work to do in redesigning our curriculum and in helping teachers change their classroom roles. Typically, in most jurisdictions, mathematics curriculum design has been about creating lists of concepts which (hopefully) match children's developmental levels. Instead, mathematics (at the levels being discussed here) needs to become contextually based so that meaning becomes part of the exploration. I am not suggesting that current 'school mathematics' be replaced with personalized routines or 'street mathematics', but I am recommending that we bring meaning to mathematics by allowing learners see its power and beauty in everyday situations. We must make mathematics meaningful by having learners discover that it connects coherently across home, school and community. Countryman (1992) puts it succinctly: 'So, here is mathematics challenging us to see ourselves and others in connection with the world' (p. 82).

Finally, we must be concerned about building students' confidence. These young women felt that they could not 'do' or even learn how to 'do' mathematics. Countryman (1992) asks, 'Is there something wrong with women or is there something wrong with mathematics?' (p. 77). I'd like to suggest that there is nothing wrong with either; instead the problem lies with the low expectations of significant others in women's lives. The expectation of many teachers, parents, and the general public is that women will be less successful in studying mathematics than their male counterparts. These negative expectations are soon 'picked up' and adopted by sensitive learners. In discussing competence in mathematics, Walkerdine (1989) comments that, 'Girls are still considered lacking when they perform well and boys are still taken to possess something even when they perform poorly.' Educators need continually to affirm women's ability in mathematics. Burton (1989) points out that:

> It is unreasonable to expect that teacher input or school environment can effect radical changes in the image of women and of mathematics in society. At the same time it is unreasonable to deny that, as a social institution, the school has considerable influence on the attitudes and behaviours of those within it (Burton, 1989, p. 185).

In conclusion, the impact of gender in learning mathematics remains an issue for concern. This study identified four factors which affect the mathematics learning of female students. By addressing each of these factors, I discovered answers to the questions posed at the outset of this study. Low self-esteem, which can have a crippling effect on the ability of some women to 'do' mathematics, evolved through the low expectations, attitudes and behaviours of teachers, parents, siblings and community members. By using a consistent and non-threatening approach, it was possible to alter the attitudes of individual female students. By changing teaching practices in the ways described, I enabled these women to make substantial gains in their ability to understand mathematics. Of course, factors which affect women as they attempt to learn mathematics affect all learners to varying degrees. I believe that when we understand the types of classroom practices which benefit females, we will have enhanced our understanding of mathematics learning in general. And then, all learners will benefit.

References

Barnes, D. (1971) 'Language and learning in the classroom', *Journal of Curriculum Studies*, 3, 1, pp. 27–38.

Best, R. (1983) *We've All Got Scars: What Boys and Girls Learn in Elementary Schools*, Bloomington, Indiana University Press.

Bishop, A. (1985) 'The social construction of meaning — a significant development for mathematics education?', *For the Learning of Mathematics*, 5, 1, pp. 24–8.

BURTON, L. (1989) 'Images of mathematics', in ERNEST, P. (Ed) *Mathematics Teaching: The State of the Art*, London, Falmer Press, pp. 180–7.

BUXTON, L. (1991) *Math Panic*, Portsmouth, NH, Heinemann.

CAZDEN, C. (1988) *Classroom Discourse: The Language of Teaching and Learning*, Portsmouth, NH, Heinemann.

COCKCROFT, W.H. (1982) *Mathematics Counts*, London, Her Majesty's Stationery Office.

COULTER, R. (1993) *Gender Socialization: New Ways, New World*, Toronto, Ministry of Education and Training.

COUNTRYMAN, J. (1992) 'Is gender an issue in math class? Perhaps it's time to change the subject', in THE NATIONAL COALITION OF GIRLS' SCHOOLS (Eds) *Math and Science for Girls*, pp. 72–85.

DESFORGES, C. (1989) 'Classroom processes and mathematical discussions: A cautionary note', in ERNEST, P. (Ed) *Mathematics Teaching: The State of the Art*, London, Falmer Press, pp. 143–50.

DEWEY, J. (1916) *Democracy and Education*, New York, Macmillan.

FULLERTON, O. (1993) An investigation of the Mathematics Register Used in Four Elementary Classrooms During Geometry Learning Experiences, Unpublished doctoral dissertation, Toronto, University of Toronto.

HALLIDAY, M.A.K. (1978) *Language as a Social Semiotic: The Social Interpretation of Language and Meaning*, London, Edward Arnold Publishers.

HOYLES, C. (1985) 'What is the point of group discussion in mathematics?', *Educational Studies in Mathematics*, 16, pp. 205–14.

LAMPERT, M. (1992) 'Practices and problems in teaching authentic mathematics', in OSER, F.K., DICK, A. and PATRY, J. (Eds) *Effective and Responsible Teaching: The New Synthesis*, San Francisco, Jossey-Bass Publishers.

MINISTRY OF EDUCATION FOR ONTARIO (1990) Mathematics Grade 6: A Report for Educators, Toronto, Queen's Printer for Ontario.

O'BRIEN, M.K. and RICHARDSON, A. (1987) *'That's the Only Maths Today.' 'Yahoo!' 'You Beauty!': Qualitative Observations of Grade 4 Mathematics Classes*, ERIC, 299 110.

PIMM, D. (1987) *Speaking Mathematically*, London, Routledge and Kegan Paul.

PIMM, D. (1991) 'Communicating mathematically', in DURKIN, D. and SHIRE, B. (Eds) *Language in Mathematical Education: Research and Practice*, Buckingham, Open University Press, pp. 17–24.

RESNICK, L. (1987) 'Learning in school and out', *Educational Researcher*, 16, pp. 13–20.

SADKER, M. (1984) *Failing at Fairness: How America's Schools Cheat Girls*, New York, Scribners.

SCHOENFELD, A.H. (1988) 'Problem solving in context', in CHARLES, R. and SILVER, E. (Eds) *The Teaching and Assessing of Mathematical Problem Solving*, Reston, VA, National Council of Teachers of Mathematics, pp. 82–92.

VYGOTSKY, L.S. (1978) *Mind in Society: The Development of Higher Psychological Processes*, Cambridge, MA, Harvard University Press.

WALKERDINE, V. (1989) *Counting Girls Out*, London, Virago.

Chapter 5

Assisting Women to Complete Graduate Degrees[1]

Lynn Friedman

Introduction

In the past five years, those concerned with gender equity in mathematics have witnessed an increasingly bright picture. In many western countries, programmes designed to appeal to women have appeared and been publicized (see reports in Burton, 1990). In the US, meta-analyses of journal and other published accounts indicate a decrease in the size of gender differences in mathematical achievement reported (Friedman, 1989; Hyde, Fennema and Lamon, 1990). A survey of new doctorates in the US reported a gradual increase in the proportion of degrees granted to women from 6 per cent in 1950 to 24 per cent in 1993 (American Mathematical Society, 1993). The first president of the AWM (Association for Women in Mathematics) gave a glowing account of the growth of that organization over the past twenty years (Blum, 1991).

Mathematics is an addictive occupation, or preoccupation, regardless of gender. Becker has described the features of mathematics which draw both women and men to the subject: 'its logical nature, its problem-solving aspects, its objectivity, and its creative nature' (Becker, 1990, p. 120). In the mid-1960s, when I completed my degree in mathematics, I felt that the support women who were considering mathematical careers needed most was company in guilt — guilt occasioned by their avoidance of personal and communal social responsibilities. But, securing company in guilt meant being able to find other female mathematicians, and there were fewer around then than there are now.

Undoubtedly, the situation has improved. This is certainly true in the western world where women faculty are becoming more numerous and increasingly visible in mathematics departments. However, they are still not well-represented at the higher levels of the profession. The proportion of women receiving graduate degrees in mathematics is still too small. In the US, gender imbalance is no longer evident at the undergraduate level. The proportion of women receiving bachelor degrees in mathematics climbed from 23 per cent in 1950 to 37 per cent in 1970 and then to 47 per cent in 1992. Table 5.1 gives statistics on bachelor degrees awarded in the period 1949 to 1992.

Table 5.1: Bachelor degrees awarded in mathematics to women, 1949–92

Year	Total	Women	% Women
1949–50	6382	1440	23
1959–60	11399	3106	27
1969–70	27442	10265	37
1974–75	18181	7595	42
1979–80	11378	4816	42
1984–85	15146	6982	46
1989–90	14597	6785	46
1990–91	14661	6917	47
1991–92	14783	6895	47

Note: Vance Grant (private communication, 1993) provided the 1990–1 figures; Norman Brandt (private communication, 1994) provided the 1991–2 figures.
Source: US Department of Education, Office of Educational Research and Improvement. (1989) *Digest of Education Statistics, Twenty-fifth Edition*, Washington, DC, US Government Printing Office, Table 266, p. 281; US Department of Education, National Center for Education Statistics. (1991) Table 233.

Table 5.2: Masters degrees awarded in mathematics to women, 1949–92

Year	Total	Women	% Women
1949–50	974	190	20
1959–60	1757	335	19
1969–70	5636	1670	30
1974–75	4327	1422	33
1979–80	2860	1032	36
1984–85	2882	1008	35
1989–90	3667	1472	40
1990–91	3615	1478	41
1991–92	4011	1559	39

Note: Vance Grant (private communication, 1993) provided the 1990–1 figures; Norman Brandt (private communication, 1994) provided the 1991–2 figures.
Source: US Department of Education, Office of Educational Research and Improvement. (1989) *Digest of Education Statistics, Twenty-fifth Edition*, Washington, DC, US Government Printing Office, Table 266, p. 281; US Department of Education, National Center for Education Statistics. (1991) Table 233.

The proportion of women receiving masters degrees in mathematics has similarly increased over the years. As is apparent from Table 5.2, 39 per cent of these degrees were awarded to women in 1992.

The figures which are cause for concern, however, are those for doctoral degrees. Table 5.3 conveys this information and shows that the percentage of female recipients has increased over the last forty years, from 6 per cent in 1950, to 21 per cent in 1992. Furthermore, according to a survey conducted by the major mathematical organizations in the country, in 1993 the figure stood at 24 per cent (American Mathematical Society, 1993). However, even the latest figures are not satisfactory. According to Henrion (1991), women make up less than 6 per cent of tenured faculty in mathematics.

Table 5.3: Doctoral degrees in mathematics awarded to women, 1949–92

Year	Total	Women	% Women
1949–50	160	9	6
1959–60	303	18	6
1969–70	1236	96	8
1974–75	975	110	11
1979–80	724	100	14
1984–85	699	109	16
1989–90	915	169	18
1990–91	978	188	19
1991–92	1082	231	21

Note: Vance Grant (private communication, 1993) provided the 1990–1 figures; Norman Brandt (private communication, 1994) provided the 1991–2 figures.
Source: US Department of Education, Office of Educational Research and Improvement. (1989) *Digest of Education Statistics, Twenty-fifth Edition*, Washington, DC, US Government Printing Office, Table 266, p. 281; US Department of Education, National Center for Education Statistics. (1991) Table 233.

Table 5.4: SAT-M effect sizes and ratios of males to females in the upper score ranges, 1981–90

Year	Effect Size Overall	Ratio Upper 0.05	Ratio Upper 0.01	Ratio Upper 0.005
1981	0.43	2.67	3.40	7.41
1982	0.44	2.76	3.58	8.14
1983	0.41	2.57	3.29	6.99
1984	0.39	2.48	3.12	6.57
1985	0.40	2.48	3.11	6.31
1986	0.42	2.64	3.40	7.47
1987	0.39	2.55	3.28	7.26
1988	0.37	2.33	2.90	5.75
1989	0.39	2.53	3.25	7.17
1990	0.37	2.38	2.99	6.19

Note: Data for this table was calculated from numbers of students taking the SAT-M, means and standard deviations, as reported by the College Entrance Examination Board Research Department, in private communications, dated 1989, 1991, and 1992.

The rise in the number of mathematics degrees granted to women is surely cause for optimism. However, researchers dealing with highly selected élite populations, such as gifted junior and college-bound senior high-school students, point to figures which are not quite so encouraging. For example, for Stanley and Benbow's 'mathematically precocious youth', the ratio of males to females is large. These individuals were tested using the mathematics section of the SAT-M (Scholastic Aptitude Test), a US college entrance examination on which scores can go as high as 800. For these gifted samples, the higher the group cut-off score, the larger the ratio of males to females, reaching 12 to 1 for scores over 700 (Stanley and Benbow, 1982). These figures are similar to those for college-bound seniors (i.e., final year secondary-school students planning to attend college). Table 5.4 gives statistics for SAT-M scores from 1981 to 1990 and includes the ratio of males to females in each of the 0.05, 0.01, and 0.005 upper score ranges of the distributions.

Ratios in the upper score ranges do not reflect the effect sizes (that is, the overall standardized mean differences) between males and females on the mathematics portion of the SAT. These have recently decreased to slightly more than one-third a pooled standard deviation. However, SAT-M male variances are substantially higher than female variances. In this case, a small mean difference, even one in favour of females, can lead to large male to female ratios at the upper ends of the distributions (see Becker and Hedges, 1988). It is true that since the mid-1970s, the ratios of Stanley and Benbow's (1982) gifted population in the upper score ranges have decreased. Moreover, the percentage of women in many other mathematically demanding professions such as computer science, engineering, and the physical sciences, is smaller than that in pure mathematics. However, those of us who are concerned with gender equity may take little consolation in this.

Opinion is divided on whether women entering doctoral programmes in the US drop out at higher rates than do men. Lewis (1991), while noting that concrete data is difficult to obtain, states that '. . . women constitute only about 30 per cent of those pursuing a curriculum that leads directly to a doctoral programme. Thus, the fact that women constitute about 25 per cent of US citizens earning a mathematics doctorate would indicate we are not losing many well qualified women at the doctoral level and that to increase the number of women doctorates requires getting them into appropriate undergraduate programs' (Lewis, 1991, p. 722). Other writers disagree (for example, Billard, 1991; Case and Blackwelder, 1992). Most of the evidence cited is anecdotal, although Billard reports a 1983 study, conducted by Berg and Ferber at the University of Illinois, which found a higher attrition rate for women despite gender-equitable distribution of financial aid. Research dealing with this question is in progress (see Case and Blackwelder, 1992).

This, then, is the problem: women receive a small proportion of the highest degrees awarded in mathematics, and possibly drop out of graduate programmes at higher rates than do men. The question therefore becomes, what does the research tell us about how to attract and retain women in these programmes? Qualitative research, in the form of interviews and case studies, has provided many suggestions. Some of the findings, in fact, are reminiscent of those confirmed in large studies of younger students. Helson, cited in Becker (1990), found that women mathematicians were less assertive, less confident, and less comfortable in their social surroundings than were their male counterparts. Becker, herself, observed that female graduate students had more doubts about attending graduate school than did men. Confidence in ability, assurance in pursuing an advanced degree, and comfort in the community of mathematicians are clearly areas that must be encouraged.

Confidence and Preparation, Interacting Factors

The finding documented most repeatedly in studies of younger students is that women have less confidence in their mathematical skills than do men (for example,

Benbow, 1988; Sherman and Fennema, 1977). In fact, confidence is a central factor in women's participation in mathematics at all levels beyond elementary school (Armstrong, 1985). Recent reports indicate that women are still participating in the most advanced courses in lower numbers than men, although there are indications that the situation is changing (Frazier-Kouassi *et al.*, 1992).

Weak preparation for university compounds the problem. Researchers have found that women tend to attribute failures to internal causes and success to external ones, while men do the opposite (see, for example, the discussion in Kloosterman, 1990). As such, women may blame their abilities for problems which are caused by their inadequate preparation. Worse still, according to Harrison (1991), 'peers and professors probably will make the same judgments due to the same social pressures' (p. 732). These circumstances can only function to diminish confidence.

Regardless of their preparation, women do comprise nearly half of the undergraduate students in mathematics. However, confidence may once again be a factor in their seemingly less ambitious choices of focus within the field (Frazier-Kouassi *et al.*, 1992). Furthermore, women graduate students often make the unfortunate decision of beginning in a masters programme to see how they like graduate studies instead of enrolling directly in a doctoral programme, not realizing that by doing so they may be labelled as 'not serious' about doing mathematics research (Harrison, 1991).

It is possible that encouragement from teachers and other experts in the field might offset lack of confidence. In Becker's study, 'the women were more influenced by "significant others" than were the men (or at least they were willing to admit it)' (Becker, 1990, p. 126). However, women are apparently less likely to find encouragement. In their undergraduate samples, Frazier-Kouassi *et al.* (1992) observed that males received encouragement more often than females to take challenging mathematics and physics courses. However, persistent personal encouragement also has its limitations. Seymour (1992) cautions that women, who tend to 'perform for others' more pervasively than do men, may be less able to handle the transfer to a more impersonal environment.

Nonetheless, sensitive and accessible advising is crucial. The problem is it is not always available. '... [W]omen and minorities are frequently inadequately counselled, and ... if anything, they require more counselling than other students to combat the gender and racial-stereotyping they receive.... [T]he need for better advising at the post-secondary level has become a serious problem' (Keith, 1989, p. 21). Moreover, mathematics faculty frequently believe that the only advice a graduate student needs is from the thesis advisor. 'For advising to be successful, its importance must be recognized' (National Research Council, United States, 1992, p. 35).

Counselling and help with academic problems should be accessible to all students on a regular and frequent basis. In addition, it should be available from mathematics faculty within mathematics departments. Women's groups or special offices created to aid women and minorities are not adequately equipped to guide and support women in specialized academic programmes.

Advisers and counsellors must be made aware of women's possible lack of confidence. Women must be encouraged to tackle ambitious programmes, and they must be provided with opportunities in lower-level graduate courses to rectify inadequate preparation. Faculty in mathematics departments must be sensitized to the need for a well-structured and supportive advising programme catering to all graduate students, but especially to the particular requirements of women.

The Social Environment and the 'Critical Mass'

The discomfort women experience in the social environment of mathematics departments ranges from feelings of being viewed as annoying intruders to simply being ignored. Social interaction between graduate students may be centred on sports in which the female student has no interest. Worse, there may be no ongoing social interaction at all, since faculty and advanced graduate students may be distant and engrossed in their own research. Such isolation is clearly diminished when there are other female students around. Harrison (1991) notes that '[t]hese problems [of male–female interaction] diminish when there are enough women around. The men are more accustomed to their presence and the women have each other for support' (p. 732). Frazier-Kouassi *et al.* (1992) cite evidence that 'women are more likely to . . . remain in a nontraditional job or programme if there are a reasonable number of other females present as classmates and/or as colleagues' (p. 49). Jackson (1992) found that one important characteristic of departments that are top producers of women mathematics doctorates is that they have women on the faculty. Female liberal arts colleges provide a disproportionate number of students who enter and complete graduate programmes in mathematics (Sharpe, 1992).

The proportion of females that might be considered adequate to ensure comfort in the mathematical milieu, that is provide a 'critical mass', probably varies from institution to institution. However, in the interests of assisting women to find graduate programmes with a 'critical mass' of female involvement, information should be made available on the number of women faculty and graduate students in each institution. In 1976, Herstein (as described in Schafer, 1991) reported on departments that had produced female doctorates. More recently, Jackson (1991) has provided an account of the top producers of women mathematics doctorates but, considering the time that it takes to produce a doctoral thesis, some universities may be doing better now than those on Jackson's list. A complete listing, by institution, of US doctoral degrees granted in 1992–3 has been published in the November 1993 issue of the *Notices of the American Mathematical Society.*

Financial Support

The western world has produced a substantial amount of legislation intended to eliminate barriers to women entering degree and professional programmes. Much of this legislation revolves around money, either by prohibiting the use of government resources by establishments which do not comply with equity standards, or by awarding grants directly to equity projects. However, early expectations concerning the levels of funding which would be provided have rarely been met, and in many cases funding levels have remained the same or even declined since the early 1980s (Stromquist, 1993). Indeed, 'Gender is an area particularly vulnerable to symbolic politics . . . the rhetoric surpassed funding and implementation' (Ibid., p. 399). Unfortunately, as a collective, women mathematicians do not yet have the political strength required to campaign effectively for better levels of support. One reason for this may be that their efforts are dispersed among several groups within a variety of larger professional associations, and in contrast to their parent associations, there is little sharing of information and combining of effort among them.

Delivery of Instruction

Apart from research on confidence and participation, our information is sparse in the area of the environmental factors affecting women in mathematics programmes. For the past forty years, there has been some focus on cognitive factors which interact with mathematical skill. In the late 1950s, the Catholic University of America produced several doctoral theses analyzing gender differences in mathematics. More specifically, it was reported that spatial and verbal skills interact with mathematical achievement in different ways for each gender. In fact, one of these researchers was prompted to suggest that since girls and boys evidently learned mathematics in different ways, they should be taught by drawing on different cognitive strengths and possibly in different classrooms (Emm, 1959). More recently, other researchers have found that females who attend single-sex schools attain better results in mathematics than do those in coeducational environments (see, for example, Lee and Bryk, 1986). However, evidence distilled from studies since the 1950s does not support the notion that males and females have differences in cognitive learning patterns except, possibly, in academically elite samples (Friedman, 1994).

It has been suggested that cooperative learning environments may be more successful in nourishing women's mathematical growth (for example, Barnes and Coupland, 1990; Frazier-Kouassi *et al.*, 1992). Other research, however, indicates that such an approach must proceed with caution (Webb, 1984). Women in small groups, where they are outnumbered by men, may find their voices drowned out and their confidence thus further diminished. Frazier-Kouassi *et al.* recommend an emphasis on problem-solving, model-building, and the discovery approach to teaching. However, research supporting the effectiveness of these approaches for females is scant.

Should we wait for the evidence? Authors of the University of Michigan project believe that 'we cannot afford to wait for research results before beginning to implement institutional change' (Frazier-Kouassi *et al.*, 1992, p. 131). On the other hand, Seymour (1992) states that '... some commonly expressed, but untested, theories about the causes of switching [out of mathematics and science undergraduate majors] have limited validity, and may serve to divert attention away from effective institutional interventions' (Seymour, 1992, p. 292). Apparently, doubts as to the theoretical viability and practical effectiveness of potential interventions was one reason for the 'cool reception' given to a group of university researchers by funding sources (Lewis, 1991). Of course without attempts at reform, the necessary research will never exist. Supporting many small-scale intervention projects may be more effective in the long run than pressing for universal methodological change.

Conclusion

The logical, objective, and non-verbal nature of mathematics, opening new worlds of reasoning to those who like to use reason, has captivated most of the readers and writers of this collection. Mathematics as it is, or was, appeals to women. Moreover, there is the emergence of a new kind of mathematics, which may be even more interesting and viable for females, in the application of mathematical modelling to areas such as the environment and public policy. We need to reinforce the strategies of the past while simultaneously searching for new methods and we can do both.

The women entering mathematics graduate programmes today are still often ambivalent or uncertain. The experience of women today is not so different from the experience of women of the 1960s as I remember it — women still feel an obligation to make a contribution to society, and many are unaware of the social value of mathematics.

We must make continued efforts to encourage change in graduate mathematics programmes: we must press for improvement in the collection and dissemination of information about female enrolment, and in the quality and quantity of the advising offered to women. Sensitive advising and the presence of critical masses of women interact to benefit female graduate students by reinforcing their confidence in their own abilities, promoting their social comfort, and enabling them to exchange information about the social value of mathematics. Finally, we must find ways to join forces by integrating rather than dissipating our efforts through the many organizations concerned with gender equity in mathematics. Only then can we apply our combined strengths to achieving these shared goals.

Note

1 The author is grateful to Carol Wood, former president of the Association for Women in Mathematics of the United States, for many helpful suggestions during the time this work was in preparation.

References

AMERICAN MATHEMATICAL SOCIETY (1993) '1993 annual AMS-IMS-MAA survey', *Notices of the American Mathematical Society*, 40, pp. 1164–99.

ARMSTRONG, J.M. (1985) 'A national assessment of participation and achievement of women in mathematics', in CHIPMAN, S.F., BRUSH, L.R. and WILSON, D.M. (Eds) *Women and Mathematics: Balancing the Equation*, Hillsdale, NJ, Lawrence Erlbaum Associates, pp. 59–94.

BARNES, M. and COUPLAND, M. (1990) 'Humanizing calculus: A case study in curriculum development', in BURTON, L. (Ed) *Gender and Mathematics: An International Perspective*, England, Cassell Educational Limited, pp. 72–9.

BECKER, J.R. (1990) 'Graduate education in the mathematical sciences: Factors influencing women and men', in BURTON, L. (Ed) *Gender and Mathematics: An International Perspective*, England, Cassell Educational Limited, pp. 119–30.

BECKER, B.J. and HEDGES, L.V. (1988) 'The effects of selection and variability in studies of gender differences', *Behavioral and Brain Sciences*, 11, pp. 183–4.

BENBOW, C.P. (1988) 'Sex differences in mathematical reasoning ability in intellectually talented preadolescents: Their nature, effects, and possible causes', *Behavioral and Brain Sciences*, 11, pp. 169–32.

BILLARD, L. (1991) 'The past, present, and future of academic women in the mathematical sciences', *Notices of the American Mathematical Society*, 38, pp. 707–13.

BLUM, L. (1991) 'A brief history of the Association for Women in Mathematics: The presidents' perspectives', *Notices of the American Mathematical Society*, 38, pp. 738–54.

BURTON, L. (Ed) (1990) *Gender and Mathematics: An International Perspective*, England, Cassell Educational Limited.

CASE, B.A. and BLACKWELDER, M.A. (1992) 'The graduate student cohort, doctoral department expectations, and teaching preparation', *Notices of the American Mathematical Society*, 39, pp. 412–18.

EMM, M.E. (1959) *A Factorial Study of the Problem-Solving Ability of Fifth-Grade Boys*, Doctoral dissertation, Washington, DC, The Catholic University of America Press.

FRAZIER-KOUASSI, S., MALANCHUK, O., SHURE, P., BURKAM, D., GURIN, P., HOLLENSHEAD, C., LEWIS, D.J., SOELLNER-YOUNCE, P., NEAL, H. and DAVIS, C. (1992) *Women in Mathematics and Physics: Inhibitors and Enhancers*, Ann Arbor, University of Michigan, Center for the Education of Women.

FRIEDMAN, L. (1989) 'Mathematics and the gender gap: A meta-analysis of recent studies on sex differences in mathematical tasks', *Review of Educational Research*, 59, pp. 185–213.

FRIEDMAN, L. (1994) 'The role of spatial skill in gender difference in mathematics: Meta-analytic evidence', Paper presented at the annual meeting of the American Educational Research Society, New Orleans.

HARRISON, J. (1991) 'The Escher Staircase', *Notices of the American Mathematical Society*, 38, pp. 730–4.

HENRION, C. (1991) 'Merging and emerging lives: Women in mathematics', *Notices of the American Mathematical Society*, 38, pp. 725–9.

HYDE, J.S., FENNEMA, E. and LAMON, S.J. (1990) 'Gender differences in mathematics performance: A meta-analysis', *Psychological Bulletin*, 106, pp. 139–55.

JACKSON, A. (1991) 'Top producers of women mathematics doctorates', *Notices of the American Mathematical Society*, 38, pp. 715–20.

KEITH, S.Z. (1989) 'The conference on women in mathematics and the sciences', in KEITH, S.Z. and KEITH, P. (Eds) *Proceedings of the National Conference on Women in Mathematics and the Sciences*, St. Cloud, Minnesota, pp. 1–5.

KLOOSTERMAN, P. (1990) 'Attributions, performance following failure, and motivation in mathematics', in FENNEMA, E. and LEDER, G.C. (Eds) *Mathematics and Gender*, New York, Teachers College Press, pp. 96–127.

LEE, V.E. and BRYK, A.S. (1986) 'Effects of single-sex secondary schools on student achievement and attitudes', *Journal of Educational Psychology*, 78, pp. 381–95.

LEWIS, D.J. (1991) 'Mathematics and women: The undergraduate school and pipeline', *Notices of the American Mathematical Society*, 38, pp. 721–3.

NATIONAL RESEARCH COUNCIL, UNITED STATES (1992) *Educating Mathematical Scientists: Doctoral Study and the Postdoctoral Experience in the United States*, Washington, DC, National Academy Press. (Available from National Academy Press, 2101 Constitution Avenue, NW, Washington DC, 20418.)

RAMIST, L. and ARBEITER, S. (1986) *Profiles, College-Bound Seniors, 1985*, New York, College Entrance Examination Board.

SCHAFER, A.T. (1991) 'Mathematics and women: Perspectives and progress', *Notices of the American Mathematical Society*, 38, pp. 735–7.

SEYMOUR, E. (1992) 'Undergraduate problems with teaching and advising in SME majors — explaining gender differences in attrition rates', *Journal of College Science Teaching*, 21, pp. 284–92.

SHARPE, N.R. (1992) 'Pipeline from small colleges' [letter to the editor], *Notices of the American Mathematical Society*, 39, p. 3.

SHERMAN, J. and FENNEMA, E. (1977) 'The study of mathematics by girls and boys: Related variables', *American Educational Research Journal*, 14, pp. 159–68.

STANLEY, J. and BENBOW, C. (1982) 'Huge sex ratios at upper end' [Comment], *American Psychologist*, 37, p. 972.

STROMQUIST, N.P. (1993) 'Sex equity legislation in education: The state as promoter of women's rights', *Review of Educational Research*, 63, pp. 379–407.

WEBB, N.M. (1984) 'Sex differences in interaction and achievement in cooperative small groups', *Journal of Educational Psychology*, 76, pp. 33–44.

Chapter 6

A National Network of Women: Why, How, and For What?

Barbro Grevholm

This chapter discusses the origins and goals of the Swedish Women and Mathematics network and outlines its connections with the society within which it operates. Through conferences and newsletters, women support each other, share information and collaborate on joint projects. Published conference proceedings in turn provide the literature for pre-service and in-service teacher education and in this way the ideas of Women and Mathematics are more widely disseminated. As well, we are engaged in research on classroom interaction and curriculum development, and other activities which promote women in post-secondary education.

Why the Need for a National Network?

Sweden is popularly regarded as a country which has made great strides in achieving equality between men and women. But, as we shall see, in the field of advanced mathematics this is still far from being realized. The Swedish educational system offers equal access to all students, providing nine years of compulsory education, and an additional optional three years at the upper-secondary school level, which the vast majority of students choose to do. It is generally agreed that girls' performance in mathematics throughout the compulsory years of education is better than that of boys, despite the fact that boys obtain slightly better results on average than girls in the national standardized test given in year nine (Grevholm and Nilsson, 1994). This picture is very similar to that found in many other studies of gender differences in mathematics achievement (Kimball, 1989).

The upper-secondary level is where different gender patterns of participation in mathematics begin to be apparent. Less than one-fifth of all students choose to pursue the natural science programme, which is a requirement for those intending to go on to university to study mathematics or science. Of this group only one-third are girls. However, they are often very successful and leave upper-secondary school with higher marks in mathematics than boys. Indeed, according to Ljung

(1990), as a group, girls who choose to study natural science enter upper-secondary school with better marks than boys, and obtain better results than boys on the national standardized mathematics test taken in the third year of upper-secondary school.

At the university level, this trend continues. Fewer than one-third of all students in mathematics are women (although 40 per cent of students preparing to teach secondary-school mathematics are women). However, at the University of Lund, Mathematics Department statistics show that while 55 per cent of male students finish their programme in the expected period of time, the corresponding figure for women is 60 per cent. Nonetheless, very few women go on to do doctoral work in mathematics. The first woman in Sweden to obtain a doctoral degree in mathematics was Louise Petrén who received her degree from Lund University in 1911. Since then, Swedish universities have granted only twenty-one mathematics doctorates (or equivalent degrees) to women (Grevholm, 1994). And according to Piene (1994), Sweden's record in this regard is worse than all other Scandinavian countries.

This poor record of female participation in mathematics is further reflected in the fact that only 3 per cent of senior university mathematics lecturers are women and Sweden currently has no female professor of mathematics. Indeed, Sweden's *only* female professor of mathematics was Sonja Kovalevskaja who died in 1891. In contrast, in other disciplines, on average about 8 per cent of all professors and 20 per cent of all senior lecturers are women (Wittenmark, 1993).

Clearly, there is a need to increase female participation in mathematics both at the upper-secondary level and at the university level. There appear to be two crucial stages in a girl's life where major barriers to participating in mathematics are experienced: the first is the point of choosing a programme of study in upper-secondary school, and the second is the passage to graduate studies. Evidently, systems change very slowly if they are not pressured from outside — there has been little change in twenty years in the field of mathematics, while in medicine and law the rate of participation of women has increased considerably over that period to the point that it is now almost 50 per cent. To change the prevailing pattern in mathematics education intervention in the educational system is essential.

The Swedish network of women seeks to increase the number of females in mathematics by engaging them in various kinds of intervention. From a theoretical point of view, the network is an intervention project, but underlying our work are a variety of feminist assumptions (see Mura, Chapter 19, this volume). According to Fennema (1994), personal beliefs and perceptions, as well as methodology, including the scientific method, are all influenced by male perspectives. Because female perspectives have not been included in a variety of areas, our view of the world is interpreted through male eyes and is incomplete at best. The ways in which women are invisible and (made) passive can be seen very clearly in various mathematics organizations of Sweden. The Swedish Mathematics Society, for example, has very few female members and only one woman on its board. Teachers' associations in mathematics and science are male-dominated even though there are many female teachers in these subject areas. For instance, the eleven-person organizing committee for the 1993 Mathematikbiennalen, a biannual national conference for mathematics teachers, had only two women members. And although most of the teachers at the younger age levels are female, only forty-nine of the 114 speakers were women. However, at the same time, it is encouraging to note that the situation is better now than it was up to 1990. As another example of the way in which women are made invisible, Swedish mathematics textbooks paint a picture, through their choice of illustrations and problem contexts, of a world that is two-thirds full of men (Areskoug and Grevholm, 1987; Rönnbäck, 1992).

Why then have we chosen to establish a network consisting only of women? Why not a network for all mathematicians and all teachers of mathematics? Do women need ways to reach each other and handle problems away from men? Projects in which boys and girls are taught separately are currently underway and members of the network are examining the effects of such interventions (Rönnbäck, 1992). Some researchers (see, for example, Hanna, 1994) have warned about the negative effects of segregated teaching. Undoubtedly, questions concerning the advantages and disadvantages of segregation are worthy of deeper analysis, but for the moment, at least, many women seem to value working in a segregated environment.

How Are We Organized?

There are of course many ways we might have chosen to organize ourselves in pursuit of our goal to empower women in mathematics. What are the advantages of a network? A formal organization would certainly offer opportunities for women to meet and make contact, but it would also require considerable organizational effort which would divert energy from other pursuits. A network offers the same opportunities without the need for special formal arrangements. A network empowers women by giving them visibility, helping them to make personal contacts, opening up communication, and making it possible for them to explore each other's work and results. It provides the inspiration and motivation for women to combine their strengths and work together to achieve a common purpose. The aims of the Women and Mathematics network in Sweden are:

- to create contacts between those who are interested in issues concerning women and mathematics, both in teaching and in research;
- to disseminate information on projects and research about women and mathematics;
- to suggest speakers (preferably female) on subjects concerning women and mathematics; and
- to be a national section of the international IOWME network (Grevholm, 1991).

In the four years since the network was established, the membership has grown from 130 to about 500 persons, a growth which demonstrates very clearly

Figure 6.1: Model of connections between the Network and the surrounding society
Note: A double arrow shows that persons belong to both organizations, a single arrow indicates the direction of possible influence.

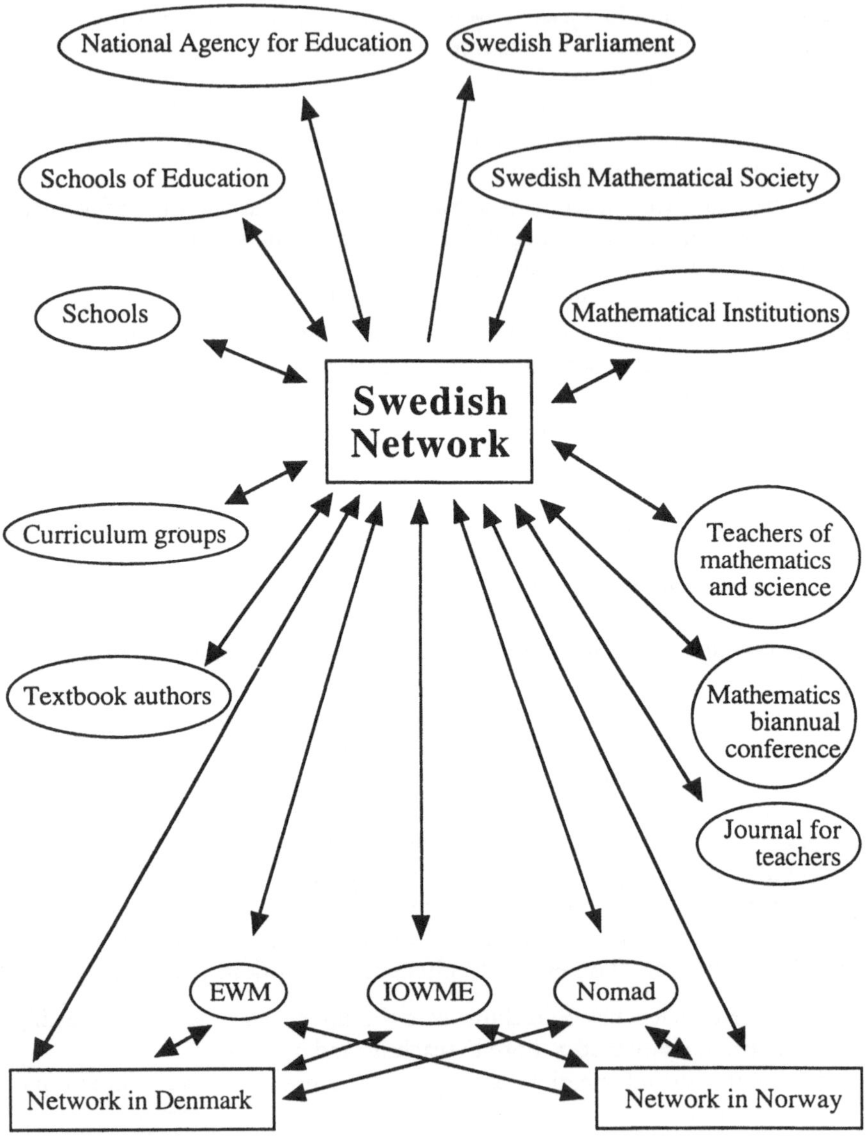

the need felt by women for finding ways to meet and discuss these issues. Connections are facilitated and information is disseminated by means of a newsletter which is distributed twice a year. Included with the national newsletter is the *Newsletter of the International Organization of Women and Mathematics*, and this ensures that news about gender and mathematics from all over the world reaches women in Sweden. Two national conferences have been organized by the members of the network and the published conference proceedings are now used as literature in teacher education (Grevholm, 1990; Brandell, 1994). The first national conference in Malmö served as a stimulus to women in the other Nordic countries: in 1991 a conference took place in Denmark (Tingleff, 1991), and in 1992 a conference in Norway was arranged (NAVF, 1993). The three networks in Sweden, Denmark and Norway have several members in common and this provides excellent opportunities for communication and the sharing of resources across countries.

Figure 6.1 illustrates the relationship between the Women and Mathematics network and other structures and organizations of Sweden and beyond. Some examples follow of the ways in which the work of the network has both influenced and been influenced by the various groups represented by the membership (for more examples, see Brandell, 1994 and Grevholm, 1992a, 1992b). Several of our members are textbook authors. As they became aware, through the network, of gender inequity in textbooks they were, in turn, better able to influence their male co-authors. The first investigation, in 1987, pointed out that boys were featured at least twice as often as girls in textbooks. An investigation in 1994 showed that a recently published textbook for upper-secondary school mathematics contains forty-three pictures of boys and men and only five of girls or women. Obviously, there is no reason to discontinue working on changing the representation of the sexes in textbooks.

Our contacts with the National Agency for Education have led to our securing financial support to hold our own conferences, to sponsor the participation of our members in international conferences, and to produce written materials on equity in mathematics education. Generous support from the Agency enabled us to host the 1993 International Commission on Mathematics Instruction's study of Gender and Mathematics Education.

Members of our network who work in schools and schools of education have helped to disseminate information about research results on gender and mathematics as well as examples of successful intervention projects in the school system. Teachers have been provided a forum for discussing the issue of whether Gender and Mathematics Education would be a suitable topic for examination in schools and teacher education.

What Are the Benefits of a Network?

Is it possible to achieve results by working together in a network? What are the outcomes for women in mathematics? Several examples have already been mentioned in the discussion above.

By creating contacts through the network, women have been able to use each other in their work as speakers, references, sources of information and inspiration. As mentioned above, women's work is often made invisible, but through the network members have been able to get to know about other women's work. For example, during the past year the doctoral theses of three women in the network have been summarized in our newsletters. Together we work for greater equality, raised consciousness about gender and mathematics, the development of the school mathematics curriculum (Johansson, 1993), and improved textbooks and other print and resource materials (Olsson, 1993).

Women must be permitted to contribute in mathematics and their perspective must be represented within the discipline and its organizing structures. There is practically no literature in Sweden about women and mathematics. The conference proceedings from the Women and Mathematics conferences are now being used as course literature by teacher education programmes at several universities. The network has put the issue of gender and mathematics education on the agenda. Through the network, women have been asked to participate and contribute to conferences (Emanuelsson *et al.*, 1992, 1994) and working groups, as well as to the international exchange of ideas. Sweden has one of the most gender-segregated labour markets in Europe (Sweden Statistics, 1994). One reason for this may be that preparation in mathematics (and science) are prerequisites for engaging in many professions. So to open doors for women to the labour market, we need first to open doors for them to mathematics.

The effect of the Women and Mathematics network is much stronger than we could have imagined when we began. It is like rings on the water after you have thrown in a stone. The results multiply and become important, not only for women, but also for men and for the discipline itself.

References

ARESKOUG, M. and GREVHOLM, B. (1987) *Matematikgranskning*, Stockholm, Statens Institut för läromedel.

BRANDELL, G. (Ed) (1994) *Kvinnor och Matematik. Konferensrapport*, Luleå, Luleå University.

EMANUELSSON, G., JOHANSSON, B., ROSÉN, B. and RYDING, R. (Eds) (1992) *Dokumentation av 7:e Matematikbiennalen*, Göteborg, Göteborgs Universitet.

EMANUELSSON, G., JOHANSSON, B., ROSÉN, B. and RYDING, R. (Eds) (1994) *Dokumentation av 8:e Matematikbiennalen*, Göteborg, Göteborgs Universitet.

FENNEMA, E. (1994) 'Mathematics, gender and research', in GREVHOLM, B. and HANNA, G., *Gender and Mathematics Education, an ICMI Study*, Lund, Lund University Press, pp. 25–42.

GREVHOLM, B. (Ed) (1990) *Kvinnor och matematik. Konferensrapport*, Malmö, Lärarhögskolan i Malmö.

GREVHOLM, B. (1991) 'Kvinnor och matematik-nätverk bildat', *Nämnaren*, 18, 2, pp. 43–4.

GREVHOLM, B. (Ed) (1992a) *Kvinnor och matematik. Rapporter om Utbildning, nr 1, 1992*, Malmö, Lärarhögskolan i Malmö.

GREVHOLM, B. (Ed) (1992b) 'Report about activities in Sweden, 1988–92', *Newsletter of the International Organization of Women in Mathematics Education*, 8, 1, pp. 14–15.

GREVHOLM, B. (1994) 'Svenska kvinnor i matematiken', in BRANDELL, G. (Ed) *Kvinnor och Matematik. Konferensrapport*, Luleå, Luleå University, pp. 62–73.

GREVHOLM, B. and NILSSON, M. (1994) 'Differential performance in mathematics at the end of compulsory schooling — a European comparison: Sweden', in BURTON, L. (Ed) *Who Counts? Assessing Mathematics in Europe*, Stoke-on-Trent, Trentham Books, pp. 241–262.

HANNA, G. (1994) 'Should girls and boys be taught differently?', in BIEHLER, R., SCHOLZ, R.W., STRÄßER, R. and WINKELMANN, B. (Eds) *Didactics of Mathematics as a Scientific Discipline*, Dordrecht, Kluwer Academic, pp. 303–15.

JOHANSSON, B. (1993) 'Utvecklingsgrupp bildad', *Nämnaren*, 20, 2, p. 34.

KIMBALL, M. (1989) 'A new perspective on women's math achievement', *Psychological Bulletin*, 105, 2, pp. 198–214.

LJUNG, G. (1990) *Centrala prov i Matematik på NT, linjerna 1985–89*, Stockholm, Primgruppen.

NAVF. (1993) *Sånn, ja. Konferenserapport*, Oslo, Norges forskningsråd.

OLSSON, G. (1993) 'Stöd och stimulansmaterial för gymnasielärare', *Nämnaren*, 20, 4, p. 12.

PIENE, R. (1994) 'Kvinnelige matematikere i Norden og i Europa — idag och i morgon', in BRANDELL, G. (Ed) *Kvinnor och Matematik Konferensrapport*, Luleå, Luleå University, pp. 133–8.

RÖNNBÄCK, I. (1992) 'Könsdifferentierad matematikundervisning i åk, 4-6', in GREVHOLM, B. (Ed) *Kvinnor och matematik. Rapporter om Utbildning, nr 1, 1992*, Malmö, Lärarhögskolan i Malmö, pp. 24–37.

SWEDEN STATISTICS (1994) *I tid och otid*, Örebro, Statistike Centralbyrån.

TINGLEFF, K. (Ed) (1991) *Kvinder og matematik, Konferencerapport*, Copenhagen, Planlägningsgruppen.

WITTENMARK, B. (1993) 'Visst är det bättre, men inte är det bra!', *Årskrönika 1992/93*, Lund, Lunds Universitet, pp. 4–5.

Chapter 7

Women Do Maths

Claudie Solar, Louise Lafortune and Hélène Kayler

Introduction

The title of this chapter derives from the title of the 1992 publication (Lafortune and Kayler, 1992) of MOIFEM (Mouvement International pour les Femmes et l'Enseignement des Mathématiques), the French-Canadian section of IOWME. Since its inception in 1985, MOIFEM has dedicated its activities to promoting women in mathematics, developing feminist approaches to the learning and teaching of mathematics (in order to attract and retain more women in mathematics and related subjects), demystifying mathematics, and establishing a French-speaking 'women and mathematics' network. Two conferences (Lafortune, 1986, 1989) were the first projects the group undertook. The conferences, designed to promote an understanding of women, mathematics, and mathematics education, generated three themes which recur in all our discussions and knowledge development: women and mathematics, the role of affect in mathematics learning, and the demystification of mathematics. The understanding we developed through these conferences encouraged us to move from theory to practice in order to tackle the issue of teaching mathematics to women. This chapter describes the development of this more practical aspect of our work.

'Maths Without Myths'

'Des maths sans les mythes' (that is, 'Maths Without Myths') is the title of the workshop we designed for women involved both in the educational system and in women's groups. Our aims were to sensitize women to the myths surrounding their mathematical abilities, to promote pedagogical approaches which are favourable to women, and to develop a network of women concerned about the status of women in mathematics. Obviously, the process of developing the objectives, content, approaches, and material, as well as the overall planning of these workshops took much time and energy. Some material was adapted from Lafortune (1990), others were developed specifically for the workshops; our feminist perspective on pedagogy emerged from our various writings (Kayler and

Lafortune, 1991; Solar, 1992) and from our group debates. Before we began working with others, we had much work to do on ourselves — it was an exciting time for creativity and the sharing of knowledge.

The publicity across Québec was very successful and it was not long before the workshops were in high demand. We targeted people who either had direct contact with girls and women through the educational system, or were playing an important role in shaping their academic or vocational orientations. The participants, a few of whom were men, were drawn in roughly equal numbers from the following groups: teachers at all levels of the educational system; counsellors, psychologists, special educators and students; students enrolled in adult-education courses, and members of women's groups and community learning centres.

Each workshop lasted a full day and the structure was adapted to suit the abilities of the facilitators as well as the needs and characteristics of the participants. The day included five activities, chosen from a bank of twenty-three, and aimed at exploring the three themes we had developed in analysing the status of women in mathematics. In the next section of this chapter we provide two examples from each of these areas. At the conclusion of each workshop, participants are provided with descriptions of all activities and asked to reflect on possible follow-up workshops they might conduct themselves within their own networks, schools, or community groups. The purpose here is to promote further network development, as well as dissemination of the activities and approaches.

The evaluations of the workshops, both formal and informal, indicate that they have been very successful. Some participants suggested that all teachers should undertake such professional development, while others said that they would use what they had learned to improve their own teaching. Still others claimed to have rediscovered mathematics and mentioned that they might go back to studying the subject. These comments, among others, have provided us with encouragement for continued work in the pursuit of equity for women in mathematics.

Sample Activities

Women and Mathematics

Une question d'image (Lafortune and Kayler, 1992, p. 108) is an activity which begins by asking participants to draw a scientist at work. Once the drawings are finished, they are displayed on the wall and each participant describes his or her creation to the group. Following this, a discussion of popular stereotypes of mathematicians and scientists ensues. Experience has shown that most drawings depict the scientist as male, although the possibility of drawing a female scientist is explicitly suggested. People usually picture the scientist as a man with no social or emotional life, no concern for politics, working alone. He is often described as cold, objective, intellectually gifted, and sloppy. Such an image is not likely to attract females to scientific careers. Holmes and Silverman (1992), in their research on Canadian teenagers, found that girls give significant consideration to their physical appearance and greatly value interpersonal relationships. Moreover, many of the drawings illustrate the scientist as having a head but no body, thus emphasizing the centrality of the mind, the separation of mind and body, and the association of mathematics and science with male intelligence.

Une histoire des mathématiciennes (Lafortune and Kayler, 1992, p. 51) is an activity that challenges the view that there are no women in mathematics. The intent here is to raise awareness of the historical absence of women, be it in mathematics or in any other field. First participants are asked to name scientists that they have heard about in the fields of mathematics, physics, chemistry or biology. After writing down the names, the small number, or total absence, of women on the list is readily apparent. To facilitate these sessions, we developed an educational poster which portrays women mathematicians of the past (Hypatia, Maria Gaetana Agnesi, Mary Fairfax-Sommerville, Sonya Kovaleskaya, Grace Chisholm Young and Emmy Noether), as well as those of the future (illustrated by five girls doing mathematics at school). The function of this representation is to help focus the discussion on the paucity of women in science, the reasons for their absence, and the manner in which society functions to exclude women from participating in mathematics. Information is then provided about the lives of several women mathematicians, stressing the difficulties they encountered in pursuing their chosen profession.

Affectivity in Mathematics Learning

Remue-méninges (Lafortune and Kayler, 1992, p. 142) begins with a brainstorming session centred on how participants feel about mathematics. The emotions are recorded and categorized as either negative or positive in nature, for instance, discomfort, fear, anxiety, frustration, and annoyance, or satisfaction, pleasure, joy, and confidence. Following this, the same strategy can be applied to specific topics of mathematics such as trigonometry, computation, probability, geometry, calculus, equations, graphs, functions or integrals. Alternatively, sheets of paper, each identifying a single topic of mathematics, can be displayed on the walls. Participants move around the room recording their feelings about each topic. The facilitator then discusses the words used and encourages a general deliberation on the role of affect and feelings in learning mathematics. This activity serves to validate the presence of emotion while learning mathematics, and reveals that the feelings involved are not all negative, nor all positive, but vary according to the individual and the specific area of the discipline involved.

J'aime, j'aime pas (Lafortune and Kayler, 1992, p. 123) is another activity dealing with affect and mathematics. Numerous strategies can be employed. One of them entails having participants fill out a questionnaire on their opinions and attitudes towards mathematics. The questions are evenly distributed among three aspects: difficulty in learning mathematics, the value of mathematics, and the pleasure of doing mathematics. Once the participants have calculated their attitude scores, they are collected anonymously in order to provide a group portrait. In this way, it becomes possible to show that there are a variety of attitudes among participants. The ensuing discussion focuses on what participants have discovered, the differences in their viewpoints, and more generally, on the training they have received in mathematics, what they like or dislike about mathematics, how useful they perceive it to be, and what areas they find easy or difficult.

Demystification of Mathematics

An example of the kind of activity we use to demystify mathematics is *Autour du cercle* (Lafortune and Kayler, 1992, p. 102), the popular method of helping students develop an understanding of π and how it relates to the circumference of the circle. (Participants measure the perimeter and diameter of the tops of various circular objects and perform appropriate calculations to obtain an approximation of π.) This activity serves to communicate the meaning of π in concrete terms. Thereafter, the participants can discuss the nature of mathematics as a science developed by human beings, its use, how it is learned, and its history. Through this process, awareness of the significance of mathematics and the meaningless way in which it is often taught are highlighted.

Another activity in this area is *Le jogging mathématique* (Lafortune and Kayler, 1992, p. 98). In contrast with those presented thus far, this particular activity is conducted over a period of time, for gradual build-up and regular repetition are essential. Its aim is to communicate the notion that learning mathematics does not require any particular gifts, rather the proper training and practice. The activity is conducted either at the beginning or at the end of each activity session. Participants are required to answer several types of questions using a definition, a symbol, an estimate, or the result of mental calculation. The questions begin simply and gradually become more difficult. At first, the workshop leader formulates the questions, but soon the women themselves can be encouraged to suggest questions. In this way, women gain confidence in both the answering and the problem-posing components of the activity. We encourage participants to keep track of their results so that they can observe improvement in their scores and thus see the benefits of practice. In conducting this activity, it is important not to emphasize speed. Emphasizing speed can place participants under great stress, as well as increase their sense of failure and lower their self-esteem. Speed is not the goal of this activity. Time limits should be used cautiously only to illustrate one possible desirable outcome of successful training. After participating in this activity, one woman stated: 'I feel more intelligent. When people talk about numbers, I panic less and less, for even though I am slower than others, I succeed in having a fair idea of what the result should be.'

Conclusion

This project has allowed us not only to develop, test and refine our workshop activities, but also to produce other types of material. For example, we have designed four pedagogical posters dealing with various aspects of mathematics: one on the 'ups and downs', the emotions involved in learning mathematics; one on women mathematicians, past and present; one on being mathematically gifted; and one on mathematical formulas and symbols related to real-life situations. In addition to these four creative endeavours, designed for use in training sessions, we produced another poster to promote women in mathematics as well as MOIFEM. It displays forty-nine photographs of women demonstrating that they participate in mathematics and evokes the image that there are infinitely many women 'doing' mathematics.

We conclude by giving a brief description of our current and proposed projects. Following the workshops, we studied the manner in which women 'do' mathematics in their lives, thus approaching a feminist critique of the discipline. We have examined how mathematics is involved in traditionally female activities such as lace-making, macramé, and quilting (Barrette and Lafortune, 1994). We have further refined our description of feminist pedagogy by presenting teachers with case studies of girls in high school and analysing their reactions (Solar, 1994). We have also re-examined the research on sex-related differences in the area of spatial visualization (Pallascio, 1994) and explored the way Inuit women use mathematics and teach it to their offspring (Paquin and Oovaut Putayuk, 1994). We have studied how mathematics is used in qualitative and quantitative research, including feminist research (Lafortune, 1994). Finally, we have used the testimony of women mathematicians to discuss their experiences with mathematical processes (Kayler and Caron, 1994). (For a detailed account of all these pursuits see Solar, 1994 and Lafortune, 1994).

References

Barrette, M. and Lafortune, L. (1994), 'La dentelle mathématique', in Solar, C. and Lafortune, L. (Eds) *Des Mathématiques Autrement*, Montréal, Les éditions du remue-ménage, pp. 115–69.

Holmes, J. and Silverman, E.L. (1992) *J'ai des Choses à Dire . . . Écoutez-moi!*, Ottawa, Conseil consultatif canadien de la situation de la femme.

Kayler, H. and Caron, R. (1994) 'Les femmes et l'activité mathématique', in Solar, C. and Lafortune, L. (Eds) *Des Mathématiques Autrement*, Montréal, Les éditions du remue-ménage.

Kayler, H. and Lafortune, L. (1991) 'Pédagogie féministe en mathématiques', *Éducation des Femmes/Women's Education*, 9, 2, pp. 27–30.

Lafortune, L. (Ed) (1986) *Femmes et Mathématique*, Montréal, Les éditions du remue-ménage.

Lafortune, L. (Ed) (1989) *Quelles Différences? Les Femmes et l'Enseignement des Mathématiques*, Montréal, Les éditions du remue-ménage.

LAFORTUNE, L. (1990) *Démythification de la Mathématique. Matériel Didactique. 'Opération boule à mythes'*, Québec, Ministère de l'enseignement supérieur et de la science, Direction générale de l'enseignement collégial.

LAFORTUNE, L. (1994) 'Femmes, recherche et mathématiques', in SOLAR, C. and LAFORTUNE, L. (Eds) *Des Mathématiques Autrement*, Montréal, Les éditions du remue-ménage, pp. 67–113.

LAFORTUNE, L. and KAYLER, H. in collaboration with BARRETTE, M., CARON, R., PAQUIN, L. and SOLAR, C. (1992) *Les Femmes Font des Maths*, Montréal, Les éditions du remue-ménage.

PALLASCIO, R. (1994) 'Visualisation spatiale et différences(?) selon les sexes', in SOLAR, C. and LAFORTUNE, L. (Eds) *Des Mathématiques Autrement*, Montréal, Les éditions du remue-ménage, pp. 171–271.

PAQUIN, L. and OOVAUT PUTAYUK, P. (1994) 'Inuttitut, mathématiques et femmes', in SOLAR, C. and LAFORTUNE, L. (Eds) *Des Mathématiques Autrement*, Montréal, Les éditions du remue-ménage, pp. 275–316.

SOLAR, C. (1992) 'Dentelle de pédagogies féministes', *Revue canadienne d'éducation/ Canadian Journal of Education*, 17, 3, pp. 264–85.

SOLAR, C. (1994) 'Femmes, mathématiques et pédagogie', in SOLAR, C. and LAFORTUNE, L. (Eds) *Des Mathématiques Autrement*, Montréal, Les éditions du remue-ménage, pp. 23–43.

SOLAR, C. and LAFORTUNE, L. (Eds) (1994) *Des Mathématiques Autrement*, Montréal, Les éditions du remue-ménage.

Chapter 8

'Girls and Computers': Making Teachers Aware of the Problems

Cornelia Niederdrenk-Felgner

Introduction

'Girls and Computers' is the title of a project, undertaken by the German Institute for Distance Education, to develop study materials for the in-service training of teachers who wish to use computers in their teaching in areas such as mathematics, writing, and social studies. At the same time, the material will be of use to teachers of a course, 'Basic Education in Information Technology', now commonly taught in all parts of Germany.[1]

Our starting point is the well-known fact that there are gender differences in how young people use and approach computers. In recent years, various studies and action research on gender issues in the use of computers have been published in Germany (see for example, Faulstich-Wieland and Dick, 1989; Heppner, Osterhoff and Schiersmann, 1990). The 'Girls and Computers' project is based on these findings with the aim of changing the classroom experience, primarily by changing teacher behaviour, in order to give more emphasis to the competencies and interests of girls.

The place of computers in the classroom is a topic of much discussion among mathematics educators. However, gender issues has been neglected in the German discussion (see, for example, Kaiser-Messmer, 1989). When computers are used in teaching mathematics or other school subjects, gender differences become very obvious, and our hope is that by discussing gender issues in relation to girls and computers, we will call attention as well to the causes of gender differences in mathematics.

Causes of Gender Differences

In the context we considered relevant for our project, the following appear to be the most important reasons for gender differences. First, there are marked differences in the ways boys and girls are brought up in Germany and, consequently, also in their adaptation to societal expectations and requirements as manifested through the education system. The development of young people is influenced in a variety of ways by sex-role stereotypes. Research with school children (Horstkemper, 1987) has shown that girls tend to be significantly less aware than boys of the potential of their abilities and skills, even though their school performance is significantly superior to that of boys. Schoolgirls who do well tend to think that they have 'just been lucky', rather than ascribing it to their own ability. When confronted with a school environment that is dominated by males and masculine values, girls tend to underestimate and give second place to their own abilities and skills, rather than demonstrate them with confidence. Males define the knowledge that is worth being taught, 'masculine' interests rate higher than 'feminine' ones, and boys receive more attention in the classroom than girls (Enders-Dragässer and Fuchs, 1989). As far as the teaching of mathematics is concerned, Srocke (1989) has demonstrated very clearly the extent to which the content and presentation of this subject have been modelled on 'masculine' orientations.

To make matters worse, teenage girls find themselves confronted with contradictory expectations. They are expected to plan their lives so as to combine the requirements of whatever occupation, vocation or profession they choose, with those of 'femininity' (Horstkemper, 1990). But the qualities that are required for success in a career (that is, the ability to achieve one's own objectives, 'rational' approaches) are not valued highly and are not accepted easily from women. In many occupations and professions, particularly those from which women have been traditionally excluded, women are now confronted with standards and norms that have been defined exclusively by men. All these factors provide an additional disadvantage to girls and young women in defining their own 'female' or 'feminine' identity (Hurrelmann, 1991).

Second, established sex-role stereotypes in German society betray a strong connection between technology and 'masculinity' on the one hand, and incompatibility between technology and 'femininity' on the other. This means that, since women are considered incapable of coping with technology and technical appliances, when it is evident that the opposite is true, such women are labelled 'not really feminine'. The relationship between women and technology is always defined in terms of failure to conform to a standard that has been set by men. A lack of skill or interest in technical or technological matters is regarded as a 'feminine deficit', competence and interest in these matters as a non-feminine deviance from an expected sex-role stereotype, whereas a more neutral view is taken in both cases for males. A widespread belief that men are the 'developers' and women the 'users' of technology prevails (Faulstich-Wieland, 1990). And this 'division of labour' adds an additional dimension to the problems already outlined.

The Problem

The 'Girls and Computers' project is very different from other German in-service teacher-training projects, both those designed to address specific school subjects, such as mathematics, and those designed to address issues, for example, pupil–teacher interaction. The problem we face in dealing with the issue of girls and computers is that most teachers are not even aware there is a problem.

Sex-role stereotypes are not confined only to students, but are also evident in the minds of teachers of both sexes and influence their behaviour. Their interactions with young people and their different expectations of boys and girls are strongly influenced by ideas on what males and females 'should' be doing. Young people, in turn, are influenced by their teachers' behaviour. They 'learn' to behave in conformity with gendered expectations. This vicious cycle can only be effectively broken by changing teacher behaviour.

There is no point, however, in simply demanding a change. Teachers have to be convinced, first of all, of the need for change. In addition they should be provided with opportunities to check, understand and modify their behaviour. This is a difficult matter to address because it is also a highly emotional one. Teacher groups that are predominantly male, as in the case of mathematics teachers, tend to avoid, or even be hostile to, confronting this question. Even the findings of serious academic research are discredited as irrelevant, with the claim that an 'abstract' subject like mathematics leaves no room for discrimination against females. On the other hand, making the observation that the same patterns of gender-specific behaviour occur between male and female teachers in teacher-training seminars, has led to interesting discussions with participants talking about issues that concern them personally.

The 'Girls and Computers' Study Material

Our teacher-training materials are designed for use in a variety of subject areas including mathematics, computer studies, German and social studies. They include a 'Basic Information' unit (Niederdrenk-Felgner, 1993) including background information on factors influencing the development of students' skills and abilities during adolescence, and other units, to be published in the next couple of years, which exemplify 'girl-friendly' lesson plans and ideas for classroom activities which involve computers, some of them accompanied by software.

The 'Basic Information' unit is based, wherever possible, on concrete examples. The idea is to sharpen teachers' awareness, and to highlight concrete situations where teachers and pupils display 'gender-specific' behaviour. Bearing in mind that teachers have their own personal beliefs and explanations, their own 'subjective theories' about problems, each chapter begins with a case study — a description of a real classroom situation, or a relevant quotation — followed by questions. In this way, teachers are encouraged to explore various aspects of the case study and to formulate appropriate strategies for taking action. A variety of questions are included to assist teachers in exploring the case studies and later sections in the text refer back to some of these questions, answering them where possible. A different type of question encourages readers to reflect on their own classroom situation and make their own observations. This may include assignments, like determining the number of male and female mathematics teachers at one's own school, or analysing school textbooks for examples of clichés and sex-role stereotypes. The idea is to enable readers, wherever possible, to make their own assessment of problems and to verify facts that have been cited in the text. Rather than being mere consumers of the text, readers are encouraged to check and seek out opportunities for change in their own schools.

Another important aspect of our methodology is to encourage teachers to take the personal experiences of their own pupils into account. The material contains sample essays for analysis. For example, essays on the topic 'A normal working day when I am 30 years old', written in test classes, show very clearly the difference between boys' and girls' perspectives in planning their futures, and the extent to which their ideas are based on clichés and stereotypes. Another essay, entitled 'What I have done with a computer. Why I enjoyed it/did not enjoy it', gives teachers an idea of the extent to which pupils in a beginners' class have computer experience. At the same time, these essays reveal the expectations, attitudes, and, in some cases, fears of girls and boys. Teachers are encouraged to ask their own pupils to write similar essays and to start similar discussions in their own classes.

In the section on classroom activities, we point out why merely reorganizing classes, for example by separating boys and girls, is no remedy in itself. However, we also point out how, at certain stages and when combined with discussion of the pros and cons of coeducation, temporary instruction in gender-separate groups may be a productive stimulus for non-stereotypical classroom behaviour.

What We Are Hoping to Achieve

The ideas on which this material is based have already been tried out in seminars for practising teachers. These were attended by teachers who had specifically applied to participate in a 'Girls and Computers' seminar although many of them were initially quite sceptical, believing, for example, that there was nothing unusual in the way males and females are presented in school textbooks, despite the obvious evidence to the contrary (men/boys appear much more frequently and in more interesting contexts than women/girls). It was only after some discussion that these participants were prepared to admit the existence of differences and the implications.

As we had expected, all participants claimed that there was no difference in the way they treated boys and girls in their own classroom, or at least that they did not intend to make any significant distinction in their student interactions. Presenting the teachers with the findings of research projects was not sufficiently convincing. What was much more effective was involving them in a classroom observation and follow-up analysis of an actual lesson that took place during the seminar. By the end of the seminar, all participants understood why the issue was important, and were prepared to reconsider their own classroom behaviour.

Our material is also intended for use in the context of vocational training of young women in technical occupations. Our ultimate hope is that this type of approach will contribute to long-term changes and improvements of attitudes.

Note

1 I wish to thank Lothar Letsche and Helga Krahn, the other members of the project team, for their helpful and constructive comments and discussions. Lothar Letsche provided the English-language version of this chapter.

References

ENDERS-DRAGÄSSER, U. and FUCHS, C. (1989) *Interaktionen der Geschlechter: Sexismusstrukturen in der Schule*, Weinheim und München, Juventa Verlag.

FAULSTICH-WIELAND, H. and DICK, A. (1989) *Mädchenbildung und neue Technologien: Abschlußbericht der wissenschaftlichen Begleitung zum hessischen Vorhaben*, Wiesbaden, Hessisches Institut für Bildungsplanung und Schulentwicklung (HIBS), Sonderreihe Heft 29.

FAULSTICH-WIELAND, H. (1990) 'Technikdistanz von Mädchen?', in HORSTKEMPER, M. and WAGNER-WINTERHAGER, L. (Eds) *Mädchen und Jungen — Männer und Frauen in der Schule. Die Deutsche Schule, 1. Beiheft*, pp. 110–25.

HEPPNER, G., OSTERHOFF, J. and SCHIERSMANN, C. (1990) *Computer? 'Interessieren tät's mich schon, aber . . .'*, Bielefeld, Kleine Verlag.

HORSTKEMPER, M. (1987) *Schule, Geschlecht und Selbstvertrauen: Eine Längsschnittstudie über Mädchensozialisation in der Schule*, Weinheim und München, Juventa Verlag.

HORSTKEMPER, M. (1990) 'Zwischen Anspruch und Selbstbescheidung — Berufs- und Lebensentwürfe von Schülerinnen,' in HORSTKEMPER, M. and WAGNER-WINTERHAGER, L. (Eds) *Mädchen und Jungen — Männer und Frauen in der Schule. Die Deutsche Schule, 1. Beiheft*, pp. 17–31.

HURRELMANN, K. (1991) 'Junge Frauen: Sensibler und selbstkritischer als junge Männer', *Pädagogik*, 7–8, pp. 58–62.

KAISER-MESSMER, G. (1989) 'Frau und Mathematik — ein verdrängtes Thema der Mathematikdidaktik', *Zentralblatt für Didaktik der Mathematik*, 21, pp. 56–66.

NIEDERDRENK-FELGNER, C. (1993) *Computer im koedukativen Unterricht: Studienbrief aus der Reihe 'Mädchen und Computer'*, Tübingen, German Institute for Distance Education at the University of Tübingen.

SROCKE, B. (1989) *Mädchen und Mathematik*, Wiesbaden, Deutscher Universitäts Verlag.

Chapter 9

Common Threads: Perceptions of Mathematics Education and the Traditional Work of Women

Mary Harris

Introduction

'Common Threads' was the name of a unique exhibition that began as a small offshoot of a mathematics education project. The result of a passing remark by a committee member, the exhibition was researched, designed, and built by the project director, almost single-handedly, in her spare time. 'Common Threads' was about mathematics, that cerebral and stereotypically masculine discipline, yet all its exhibits were needlework, that practical and stereotypically feminine pursuit. By presenting these two extremes in terms of each other, 'Common Threads' demonstrated the wealth of mathematical thinking and skill that goes on within the very practices that traditionally mark their antithesis.

The title 'Common Threads' was a deliberate play on words chosen to illustrate the shades of meaning in the exhibition's message. Thus, the word 'common' was used in the sense of *ordinary, everyday,* as well as in the sense of being *widespread.* The word 'threads' referred to the *yarns* of knitting, the *silk* of embroidery, the *grasses* of a basket, and the *warp* and *weft* of weaving. But it also implied the developmental strands of mathematics, both the vertical paths through which an area of mathematics develops in depth, and the horizontal links between various branches of the discipline. In keeping with the cloth analogy, 'Common Threads' represented a seamless robe of mathematical activity within a very ordinary part of anyone's daily life, anywhere on earth.

The exhibition was originally displayed for just one week, however, its impact during that week was such that it subsequently toured England for a period of two years. In response to international interest, two copies of a new version were later created by the British Council, and these travelled to more than twenty countries in Europe, Africa, Australasia. and America over a period of four years. During its tours, 'Common Threads' has stimulated curriculum development in school mathematics, cross-curricular work from elementary to tertiary levels, vocational work in formal and informal settings, ethno-mathematics research, and programmes in women's development.

Two Exhibitions

The new version of 'Common Threads' was substantially different from its predecessor, although the original title was retained. For the sake of clarity in this chapter, the original will be referred to as 'Common Threads 1', and the redesigned versions will be termed 'Common Threads 2'.

'Common Threads 1' was designed and created by the project 'Maths in Work' which I directed from 1981 until its completion in 1991. It was an educator's homemade exhibition, the outcome of a particular perspective in mathematics education discussed in more detail below. 'Common Threads 1' consisted of five sections: symmetry, number, creativity, information handling, and problem-solving. The symmetry section, which set the tone for mathematical continuity in breadth and depth throughout the entire exhibition, was developed mathematically through panels explaining the ways in which mathematicians think of space, including references to symmetry and transformations. It began with a piece of cross-stitch embroidery that I had purchased in a craft market while on holiday in the former Yugoslavia. I illustrated the processes whereby an embroiderer develops mathematical symmetry on cloth used like graph paper, by making a set of small embroidered panels stopped at different stages of the design development. My comments as to how I, the embroiderer, had built up the patterns formed the captions, and I placed mirrors beside the exhibit so that visitors could explore the design stages for themselves. I illustrated strip patterns using hair ribbons and pieces of upholstery trimming, and the section ended with a mathematical description of the wall-paper patterns shown on dress materials and a Turkish kilim rug. The concept of symmetry was not confined to this section, however, but overlapped with the other areas to demonstrate its significance throughout mathematics. For example, the number section, primarily illustrated through knitted objects, contained Aran and Fairisle sweaters whose symmetrical designs were analysed in terms of number groups.

'Common Threads 1' broke new ground as a learning resource, but, as I was later to discover, it also broke at least two rules of exhibition design. First, since cloth is tactile and inviting, it never occurred to me to say 'Do Not Touch'; on the contrary, I intended the materials to be touched. I also provided a table with graph paper, pencils, and other materials people find helpful when doing mathematics. In the event, nothing was damaged or stolen, indeed as the exhibition toured, people sometimes even added to it. Evidently, the exhibit was treated with affection as well as with respect, and it sometimes returned to base with artifacts carefully mended.

Second, I did not know the received wisdom on captions in the field of professional exhibition design. Since the purpose of my captions was to offer mathematical explanation of what transpired in the making of the cloth items, I wrote them of a length necessary to do just that. In order to make the captions comfortable to read, I did place text and diagrams in inviting ways, but it never occurred to me to limit their length. In the case of the symmetry section, for example, some of the captions extended to whole panels. However, their effectiveness was demonstrated by the classes who visited and revisited 'Common Threads 1' as part of their school work on transformation geometry. Members of the public also carefully referred to them, taking time to read, analyse, and understand.

The two 'Common Threads 2' exhibitions were smaller, but dramatically more beautiful, and built specifically for overseas touring by professional designers from the British Council Exhibitions Department. An immediately obvious difference between 'Common Threads 2' and the original exhibition, apart from the professionalism of the design, was that all the tactile materials were now behind perspex. More subtle, however, was that the captions of 'Common Threads 2' demonstrated a perception of mathematics education that differed from the original. Not only was the mathematics content considerably reduced on aesthetic grounds ('Common Threads 1' captions were far too long!), but the designers felt that the general public would not be able to cope with the depth of mathematics explanations originally offered. The combination of the reduced captions and the spectacular cloth the new designers produced effectively fragmented the exhibit. Thus, although there was still some overlap between the new sections of symmetry, geometry, number, and codes, overall there was much less mathematics and it was presented at a more superficial and disjointed level. Whereas 'Common Threads 1' had been an interweaving of mathematics and cloth, 'Common Threads 2' was cloth with mathematical comment. The view of mathematics as low level, disconnected, and mainly arithmetical techniques applied to various situations, is already the predominant public perception and 'Common Threads 2' was made to fit it. In addition, design considerations now took priority over educational ones, and the exhibition was no longer mine. The professional understanding of mathematics as a process which the exhibition was originally intended to convey had slipped away, and I was forced to devise support materials to enable 'Common Threads 2' to be used in the same way that I had used the original exhibition.

Work as a Resource

The project that generated 'Common Threads 1', 'Maths in Work', was designed to investigate the skills new school leavers employed in their first jobs. By means of a questionnaire survey, the idea was to use the data generated to monitor the effectiveness of a new secondary school-leavers' course, as well as a source of realistic learning materials for schools. A problem I struggled with, as project director, was how to reconcile behavioural data generated by a vocationally oriented piece of research with my constructivist philosophy of school-mathematics education. I have discussed this issue in detail elsewhere (Harris, 1991). Suffice it to say that, the experience of analysing the data, and of finding more mathematics in the answers to questions that *were not* about mathematics than in answers to questions that *were*, eventually led me to abandon the research data as a true picture of mathematics in the workplace. I replaced it with my own participatory observations which, in turn, led to a complete rethinking of the learning materials I was to produce.

A feature of the research questionnaires was that the questions generated two sorts of response, a numerical frequency score of *how often* a particular skill was used and a description of the *contexts* in which frequently used skills were used. The item about context was included as a check that the worker had understood the question. By analysing this context data in addition to the skills frequency results, it was clear that workers were using many mathematics skills in their work although they might not admit to it when asked directly. For example, the hairdressers in the sample, when asked how they used ratio and proportion in their work, denied using them. However, when asked about how often they mixed substances, they explained that it is essential to mix hair dyes in the correct proportions otherwise a customer's hair could be damaged.

Two aspects of this and of many other similar responses were significant. First, there seemed to be a general negative attitude towards school mathematics. Many of the young workers had not enjoyed their school experience of mathematics and tended to respond negatively to the questionnaire as though they viewed it as a test. Second, they often did not connect what they did at work with what they had done at school, nor did they expect to. The demonstrable understanding of proportion in a practical situation displayed by the hairdressers raised no connection in their minds with the algorithms of school mathematics. Many examples of this phenomenon led me to think that mathematics at work only becomes mathematics when workers think they cannot do it.

It was this distorting emphasis on skills as a picture of mathematics in the workplace that caused me to abandon analysing the skills data and instead to take to the workplace myself. Since negative attitudes to mathematics are so widespread, I tried not to mention the word 'mathematics' at all when asking my fellow workers about their workplace problems. People love talking about their problems and it was very interesting to note that the resolution nearly always contained some mathematical thinking, though often not of a sort included on the skills list of researchers in the area of mathematics and work.

The assumption that mathematics in the workplace at this level is wholly arithmetical is widespread. It appears, for instance, in the Cockcroft Report (1982) in the chapter on workplace mathematics, where almost every use of the word 'mathematics' actually means 'arithmetic'. Of course, underlying assumptions invariably condition the answers to questions. Actually doing the work, rather than asking questions about it from a preconceived standpoint, reveals a much broader picture.

> If you are a scaffolding rigger, you don't do any calculating while you are rigging (and wouldn't Cockcroft say therefore that riggers don't use much maths?). But if you haven't grown the calibrated eyeball that knows about stresses and load bearing and bracing and moments and estimating lengths and interpreting plans you won't last long as a rigger, nor will the person below you. (Harris, 1985, p. 44)

The same applies to cardboard box makers:

> The processes that a designer of cardboard boxes, or a designer of anything else, works through are those that a mathematical problem solver goes through. They start by researching the topic, exploring the parameters, then doodling ... before going on to choosing what seems like a good way, exploring it, refining it, rejecting it if it doesn't work and eventually coming up with something satisfactory. I would argue that these processes generalise to other problematical situations whereas the specific skills maybe don't just because they are specific ... (Harris, 1985, p. 44)

The key is to ask questions not about mathematics skills on the job, but about workplace problems, for 'by asking questions about where the gremlins usually strike, you get not only a true picture of what life is like but many examples of the problematical situations which demand the real thought rather than the routine calculations which don't' (Harris, 1985, p. 45).

As my experiences as a worker grew, I came to view job sites as 'resources' for mathematical thinking instead of 'destinations' for the application of mathematics learned in school, a complete reversal of the accepted practice. The implications of this view for the design of teaching materials were also uncompromising. At that time, learning materials for mathematics at the school-work interface were of a particular kind. Typically, they took a skill, often something that had caused pupils and teachers considerable pain over a number of years, such as the addition of fractions, and then demonstrated the use of those skills at work. The assumption was that by placing the skill in context, the pupil would see the necessity of learning it, as well as come to understand its use. The educational reality, however, was the presentation of a distorted, limited and limiting view of mathematics to a group of already disenchanted pupils.

The authors of such materials may or may not derive their inspirations from real workplaces. If they did, they tended to edit out variables so as to make the mathematical skill component of the question more salient. If they did not, they invented hypothetical workplace situations, again for the purpose of illustrating the mathematics. In either case, the materials were not a true representation of reality, and the first people to recognize this were the pupils. Since I had reversed this way of doing things, I had to find a way of working with the entire context of the workplace as a generator of problems, not as a setting for the practice of isolated skills taught separately and then applied like a bandage to a wound.

Perceptions of Mathematics Education

The first pack of learning materials produced by the 'Maths in Work' project began on the floors of cardboard-box factories and generated a set of 'design and make' problems for mathematics teachers. It was called 'Wrap it Up' and contained about twenty-five problems in a form that could not possibly be mistaken for a mathematics textbook. Each open-ended problem was presented with a minimum of instructions on a single sheet of paper. The format was such that teachers could select problems, and adapt them to the needs of their own students. In this way, the role of the teacher became one of professional educator as opposed to skills trainer. Because the problems were real, the materials did not talk down to anybody and there was no suggestion in either form or content that the work was low-level; indeed, the level was what the student wanted it to be. The pack proved so successful with teachers that I developed the model for other workplaces.

The origin of the type of research and learning materials that I came to reject lies in the social history of the British mathematics curriculum. I refer, here, to a society so obsessed with social class, that it produced entirely different sorts of mathematics education for different pupils based on their societal standing (see, for example, Williams, 1961, for a very readable account of this period). The legacy of liberal mathematics (which of course contained Euclid) for the upper classes of society, and utilitarian mathematics for the lower classes, can still be witnessed today in the disconnected skills included in the British National Curriculum and the debate that surrounded them (National Commission on Education, 1993). It is evident too in the manner in which companies recruit different groups of workers. At all levels, mathematics knowledge is still regarded as both a criterion of intelligence and a mark of the type of education a candidate has received. A company will recruit graduate mathematicians, not because it needs their mathematics, but because a mathematics degree is evidence of the 'best sort of brain'. At the bottom end, 'unskilled' jobs and the 'basic skills' of school arithmetic are virtually defined in terms of each other. In fact, the closeness of the relationship results in its being unquestioned and therefore reinforced by much of the research aimed at investigating it.

'Common Threads 1' and the English Tour

Different perceptions of mathematics education have had different effects at different times. A good review of five educational ideologies and their philosophical foundations and implications is provided in Ernest (1991). In the 1970s, increased employer concern about the low attainment of school leavers stimulated research in this area. By the mid-1980s, two primary concerns in mathematics education were centred on gender and culture. The research question at that time was 'Why are so many girls and members of ethnic minorities outperformed in school mathematics by dominant culture boys?' In seeking to address this question through the materials I was developing for my second learning pack, I looked for something that was as ordinary as the cardboard of the first pack, but was particularly resonant with girls of any culture.

Cloth is such a substance. It is so ordinary that it is taken for granted by those not in the textiles business. Everywhere in the world people wear clothes, and these are usually produced by women at home or in a factory. In fact, the productive work of women is of greater importance in the textiles industry than in any other trade (Holland, 1991). Work with textiles in all locations, and in most cultures, is thoroughly gendered. Thus, if I could demonstrate that the very medium which defines the lowly work of women worldwide is richly mathematical, then I could explore a fresh perspective on gender, culture, and mathematics education.

Work began on a resource pack using the same format as 'Wrap it Up'. All the problems were taken from the textiles workplace, and as the work progressed, the 'Maths in Work' office began to fill with interesting cloth from all over the world. It was this collection that inspired the chance remark that led to the 'Common Threads' exhibition. Research for the exhibition was an extension of the work already under way. As an experienced amateur needleworker myself, I already had practical experience regarding the mathematical basis of knitting, sewing and, weaving. Because I was also experienced in mathematics, it was a relatively simple matter to rework particular pieces of needlework using the techniques and materials of the original worker, while simultaneously bringing out the mathematical thought inherent in the design development. For example, adapting a knitting pattern to produce a garment of an intended size and shape is entirely a matter of ratio; designing a decoration for a garment is a question of geometry and number manipulation; matrices form the mathematical foundation required for weaving.

My lack of experience in the industrial sphere made research in this area both challenging and rewarding. It was here, however, that I was immediately confronted with the public perception of mathematics in 'unskilled' work as being nothing more than arithmetic. At almost every factory I visited, I was greeted with tirades against schools for not teaching long division, or by complaints that school leavers cannot multiply fractions. A visit to a tie factory was but one example and will serve to illustrate. On arrival at the factory and after the usual opening rounds, I was invited to look at some new software in the accounts department, the presumed destination of my visit. Only after a considerable amount of persuasion was I allowed to visit the cutting room. Here, I found workers manipulating the cloth from which ties were to be made, so as to cut the three pattern pieces, without wasting any cloth, in such a way that the motifs on the finished articles would be horizontal and in the correct place. In direct contrast to school mathematics in which geometry is done on sheets of paper in portrait orientation (with triangles sitting firmly on their bases), the whole business of making ties is geometry for real at the odd angles of real life, demanding sophisticated mental and physical manipulation of spatial images and artifacts. Readers are invited to unpick a necktie and investigate this for themselves.

At the opening of 'Common Threads', the enthusiasm was so great that it had to be kept in check for safety reasons. Schools had to be turned away because the gallery, which was also open to the public, was not big enough, and it was for this reason that the exhibition went on tour. Enthusiasm from members of the public was also marked. On one occasion, a child of about 4 years of age stood at the door, grabbing the hands of adults as they entered and insisting that they 'come and look at these lovely patterns'. An Indian academic, walking past the gallery on her way to a meeting, was drawn to the exhibition by the mathematical analysis of the designs on the hem of her sari.

Once on tour, enthusiasm continued. Schools, craft workers, and passing visitors worked in concert to add local colour, and a number of cross-curricular projects were taken up within the schools. In one location, some lace makers complained that there was not enough lace in the exhibition and, in response, brought their own work to the exhibition and performed live demonstrations. At the end of one working day, they left their equipment on the table despite warnings that children would touch. Their response was 'Of course. How else will they learn?'

As 'Common Threads 1' toured, it became clear that it did respond to the issues it was designed to address. It raised the confidence of female pupils and teachers by confirming the nature and extent of the mathematics in 'their' work. It confirmed the traditional narrowness of school teaching and textbooks, and exposed the gendered nature of the examples they offer. No longer was it possible to deny the richness of mathematics that exists within the cultures of children who were not achieving highly in school mathematics. The question 'Why are so many girls and members of ethnic minorities outperformed in school mathematics by dominant culture boys?' changed to 'Why is school mathematics still so narrow-minded?'

The results of the tour and this change in questioning were reported in 1988 at ICME-6 in Budapest where they attracted the attention of delegates from a number of countries, notably Australia and New Zealand. It was requests, by educators from these countries, to borrow 'Common Threads' that led the British Council to take the imaginative leap of redesigning 'Common Threads' and taking it on tour.

'Common Threads 2' and the Overseas Tour

The two copies of 'Common Threads 2' began touring at the end of 1990, one in Denmark and the other in Turkey. After Denmark, the first copy went on to Norway and then south to Nigeria. After Cameroon, Uganda, Malawi, Kenya, Zimbabwe, Botswana, Lesotho, and Swaziland, its journey concluded in Tanzania. The second copy headed south-east from Turkey, to Thailand, Malaysia, Singapore, Australia, and New Zealand, after which it crossed the Pacific to Brazil and finally Canada.

Limitations of space do not allow for a detailed elaboration of the variety of work associated with the 'Common Threads 2' tours. The ways in which it has been used have depended on both the educational policies and the perceptions of its hosts. Gender and mathematics issues are a matter of culture and vary from country to country, including those in which their existence is officially denied. When 'Common Threads 2' was provided to the mathematics education community of a particular country, it acted either as a resource for work already in progress, or as a stimulus for new investigations. For example, in Botswana, the university collaborated with the British Council in developing a series of programmes for schools, and eventually a book of cross-curricular activities based on Botswanan culture (British Council, 1993) which was distributed to every secondary school in the country. In Thailand, an exhibition of Thai textiles, complete with their own captions, was designed and set up alongside 'Common

Threads 2' in two locations, a department store and an annual conference of mathematics and science teachers. This heightened awareness of the local environment as a resource for learning, and has resulted in a textiles and mathematics display at every annual teachers' conference in Thailand since the time of the original exhibition.

Discussion

The very nature of academic work divides and isolates studies that ought to be allowed to cross fertilize. The effectively separate histories of women's education, textiles work, and mathematics education make interesting parallel reading (Alic, 1986; Boyer, 1989; Parker, 1984). Parker, in her book *The Subversive Stitch* (1984), explores and interprets the processes whereby embroidery, the highly respected medieval *art* of both men and women, was down-graded to a *craft*, hierarchically gendered through power-brokering and social control, to the point where embroidery became almost a genetic characteristic of nineteenth century femininity. The first use of embroidery in the curriculum was to differentiate girls' education from that of boys', a means whereby girls could be educated without posing a threat to boys. By the late nineteenth century, its place in the curriculum had become vocational needlework for domestic service, handwork for the nimble fingers and feeble brains of girls, and an appropriate educational alternative to the difficult sums provided for boys.

Surrounding the history of embroidery is the story of how women's work was down-graded, deskilled, underpaid, and exploited through processes of differentiation, marginalization, exclusion, segregation, and subordination (Holland, 1991). As female employment moved from the home to the factory, the sexual division of labour also transferred, ensuring that women remained hierarchically subordinate in the labour market. Within the work environment, gender hierarchy is maintained through technological advances; men maintain control over the processes and know-how, while women operate the machinery with no idea of how it works. This process of gendering produces extremes whose immorality is masked by their sheer stupidity. In one factory I visited while researching 'Common Threads 1', a room full of workers machined children's clothes swiftly and accurately, maintaining parallel lines and curves on the small pieces of cloth. In the next room, other workers picked off the fluff that accumulated on machines knitting socks. The union classification of the jobs were 'skilled' for the fluff pickers who were white men, and 'unskilled' for the machinists who were Asian women.

In the field of mathematics education, the seminal work of Walkerdine and her colleagues details the processes whereby girls are 'gendered out' of achievement in mathematics (Walkerdine, 1989). The forces that ease women and girls out of high achievement in mathematics, the forces that maintain mathematics as a male domain, the forces that stereotype particular activities as women's, and the forces that label those activities trivial and brainless are one and the same. In my own work, I have seen male mathematics teachers refuse to work with textiles in a mathematics workshop because, as 'women's work', they found it degrading. I have female colleagues in mathematics education who refuse to do needlework, not because they do not like it, but because of the connotations it has as feminine, trivial and time-wasting. Both perspectives are equally prejudiced.

Public perceptions are hard to shift. It is difficult not to see things in a gendered fashion when they are so prevalent in our culture. If the public perceive mathematics as a male subject in spite of the evidence to the contrary, then it will take at least a generation of education to change it. As Willis (1989) points out, 'Since we do not award prizes for low levels of achievement, boys' over-representation amongst the mathematically least able does not receive much media or educational coverage' (Willis, 1989, p. 14).

The damaging differences in perceptions as to what mathematics education is about, both inside and outside the profession, should be more tractable from within. A largely undiscussed problem in mathematics education is that mathematics communicates mainly with itself. In England, for example, we still have not grasped that the public's perception of mathematics education, the one that ultimately has political power over the discipline, is totally different from its own (Harris, 1989). There have been programmes attempting to popularize mathematics, television programmes of mathematical fun and games, we have even improved participation rates and grades in school, but we have achieved very little movement amongst the power groups of government and industry who control education and the curriculum and who 'are raucously engaged in demanding a different scenario from the one we are offering' (Harris, 1989, p. 18). A siege mentality within mathematics education is not the way to get things changed.

The greatest general effect of both 'Common Threads' exhibitions was that they communicated widely outside mathematics education as well as inside it. There is no doubt that they did change perceptions. Their main message, however, was that the very stereotype that currently limits women in both mathematics and work can be used as a weapon for their emancipation by the recognition, development, and accreditation of the power tool that already exists within.

References

ALIC, M. (1986) *Hypatia's Heritage: A History of Women in Science from Antiquity to the Late Nineteenth Century*, London, Women's Press.

BOYER, C.B. (1989) *A History of Mathematics*, 2nd ed., New York, Wiley.

BRITISH COUNCIL (1993) *Common Threads in Botswana*, Gabarone, British Council.

COCKCROFT, W.H. (1982) *Mathematics Counts: Report of the Committee of Inquiry into the Teaching of Mathematics in Schools*, London, Her Majesty's Stationery Office.

ERNEST, P. (1991) *The Philosophy of Mathematics Education*, London, Falmer Press.

HARRIS, M. (1985) 'Wrapping it up', *Mathematics Teaching*, 113, (December), pp. 44–5.

HARRIS, M. (1989) 'Basics', *Mathematics Teaching*, 128 (September), p. 18.

HARRIS, M. (1991) 'Looking for the maths in work', in HARRIS, M. (Ed) *Schools,*

Mathematics and Work, London, Falmer Press, pp. 132–44.

HOLLAND, J. (1991) 'The gendering of work', in HARRIS, M. (Ed) *Schools, Mathematics and Work*, London, Falmer Press, pp. 230–52.

NATIONAL COMMISSION ON EDUCATION (1993) *Learning to Succeed: A Radical Look at Education Today and a Strategy for the Future*, Report of the Paul Hamlyn National Commission on Education, London, Heinemann.

PARKER, R. (1984) *The Subversive Stitch: Embroidery and the Making of the Feminine*, London, Women's Press.

WALKERDINE, V. (1989) *Counting Girls Out*, London, Virago.

WILLIAMS, R. (1961) *Education and British Society in the Long Revolution*, Harmondsworth, Penguin Books.

WILLIS, S. (1989) *Gender and the Construction of Privilege*, Geelong, Victoria, Australia, Deakin University Press.

Part 2

The Cultural Context

The chapters in this part of the book provide a cross-cultural perspective on the discussion of equity in mathematics education. We have grouped the chapters in two sections according to whether they are based on comparative research or focus on particular issues within a single country. The chapters in the first group explore differences either among several cultures or among several ethnic groups within a single culture. The second group of chapters provide a perspective on the current situation in several countries across the world. Taken together, these papers challenge monocultural assumptions underlying previous intervention approaches and question their suitability outside the culture in which they were developed. Several also question the relevancy, given local circumstances, of maintaining a concern for equity in mathematics education.

A comparison of the cultural influences contributing to gender differences in learning mathematics within different cultures is the focus of the chapter by Gurcharn Singh Kaeley. He observes that the developing world witnesses a greater disparity in the education of both sexes than is apparent in the developed world, where gender differences favouring boys have by and large disappeared, at least in the compulsory years of schooling. By contrast, he points to the situation in certain matrilineal societies where females achieve the same or better mathematics results than their male counterparts. Hawai'i is the only state in the US in which gender differences in mathematics favour girls. Paul Brandon, Cathie Jordan and Terry Ann Higa analyse the reasons for this phenomenon by examining the different ethnic groups within Hawai'i. Several factors are suggested, among them, differing adaptation to Hawai'i by male and by female immigrants, differential school compliance, and the differential impact of peer culture among boys and girls of native Hawaiian ancestry. In the same vein, Sharleen Forbes examines gender differences in mathematics performance between the two different ethnic groups in New Zealand. She reports that the average performance of Maori girls is lower than that of the Maori boys, whereas for girls and boys of European origin the average performance is the same. It appears that strategies to increase the participation and achievement of girls in mathematics in New Zealand have had a positive impact on girls of European descent, but have not met the needs of Maori girls. This group of chapters concludes with a study, by Gilah Leder, of the influences of the print media on gender differences in learning mathematics. Focusing on print media in two ostensibly similar countries, Australia and Canada, Leder distinguishes among the media images using feminist and societal-psychological lenses. She concludes that subtle messages conveyed by the popular press are consistent with small but persistent differences in the ways females and males perceive and value mathematics.

A strong challenge to monocultural perspectives on equity in mathematics education begins the second group of chapters in this part of the book. Saleha Naghmi Habibullah presents a personal view of the issues in which she questions whether equity in mathematics education is an appropriate goal in a developing country. Even were it so, she challenges the assumption that western strategies for overcoming gender inequities would be equally effective in non-western societies. In her opinion, there is a direct link between a country's socio-cultural conditions and the manner in which its people perceive and react to the issue of gender imbalance. Consequently, for many years to come, short-term goals and strategies for educational improvement will have to vary between countries according to economic conditions and socio-cultural realities.

The remaining chapters in this section describe the situation in a single country and emphasize the influence of the cultural context on the learning of mathematics. Berinderjeet Kaur surveys the available empirical research documenting gender differences favouring boys in mathematics achievement and attitudes in Singapore. Neela Sukthankar investigates the cultural issues affecting the women's education in Papua New Guinea. She describes the disadvantages girls endure as a result of low enrolment at all school levels and attending poorly equipped schools. The main factor she cites as producing these disadvantages is family poverty which results in most families being able to send only one child to a good school; because of traditional gender-role expectations, this is usually a boy. Other contributing factors which she explores are the loss of traditional Papua New Guinean pre-mathematical concepts and language barriers to mathematics education.

Two chapters in this part of the book deal explicitly with the issue of single-sex mathematics education. Françoise Delon describes the impact on women's participation in mathematics education after the recent desegregation of the most respected universities in France. Her account details the significant decrease that has occurred in the numbers of female students studying mathematics and, ultimately, in the numbers of females pursuing careers in universities, schools and other prestigious professions. The former women-only learning environment had been reassuring and stimulating to women and empowered many of them to enter mixed competitive working environments common in the French society. Like Delon, Pat Hiddleston also advocates gender-segregated schooling, but in contrast to Delon, her account describes the situation in a very poor developing country. According to Hiddleston's research, girls in Malawi who attend girls-only schools achieve better results in the final school examinations than those who attend coeducational schools and more of them enrol for a university education than girls from coeducational schools. Furthermore, at the university level, although they enter university with lower overall and lower mathematics qualifications than males, by the time they graduate, female students have higher results in mathematics and the sciences on average than male students.

Chapter 10

Culture, Gender and Mathematics

Gurcharn Singh Kaeley

Introduction

Recent meta-analyses of gender differences in mathematics performance indicate that generalizations as to the superiority of either gender are impossible (Feingold, 1989; Friedman, 1989; Hyde, Fennema and Lamon, 1989). The magnitude and direction of such differences depend on a number of factors. For instance, age, type of task, sample selection (Hyde, Fennema and Lamon, 1989), and the nature of the evaluation procedures (Kimball, 1989) all play a role. Female students appear to demonstrate superiority over their male counterparts on computational, numerical, and perceptual-speed tasks before high-school, with little or no difference after that time. However, on problem-solving and spatial tasks, male students seem to outperform females by the time they reach 13 years of age (Lummis and Stevenson, 1990; Xu and Farrell, 1992). Boys achieve higher scores on standardized tests, as well as in the mathematics classroom (Kimball, 1989). Quoting Academic Association of University Women, Streitmatter (1994) corroborates that both the National Association of Educational Progress and College Board data in the US indicate that males outscore females on test items that assess higher cognitive skills. Overall, the best one can conclude is that there appears to be a modest difference in mathematics performance in favour of male students.

Gender differences, such as those mentioned above, stem from cultural pressures and socialization processes characteristic of many countries in which female students are not permitted to develop their mathematical abilities to their fullest potential. Beliefs and prejudices about sex-appropriate behaviour are reflected in the expectations of parents, peers, school, and society (Leder, 1985). Such cultural stereotypes not only include the alleged natural superiority of boys' mathematical abilities, but also differences in beliefs about the utility of the discipline for boys and girls. In maintaining these perspectives, societies undermine girls' confidence and motivation, and impoverish their learning of mathematics (Meece *et al.*, 1982).

This chapter investigates the relationship between culture and gender differences in mathematics enrolment and performance. In the pursuit of this goal, four hypotheses are considered. First, the cultural norms in many developing countries are responsible for producing enrolment disparities. Second, in the developed world, cultural norms operate to discourage female students in mathematics to the point that their enrolment in mathematics courses declines as soon as enrolment in the subject becomes optional. Third, in societies where the role of women has changed, gender differences in mathematics performance are beginning to decrease. Finally, in certain societies and cultural groups in which women already have more power and authority, females outperform males in mathematics.

Enrolment Disparities

Developed Countries

In North America, Europe, and many other developed countries there is universal, compulsory primary and secondary-school education provided through government taxation. In some of these regions, such as the UK, even tertiary education is subsidized with students from disadvantaged backgrounds receiving financial assistance to pay for their books and other necessities.

Many researchers have studied enrolment patterns as well as performance in mathematics courses in developed countries. There is consensus in their findings that female students enrol in smaller numbers than boys in the more advanced optional high-school and college mathematics courses. Such reports derive from a wide variety of countries such as the UK (Shuard, 1982), the US (Fennema, 1985), Australia (Leder, 1980) and a majority of other countries included in the Second International Mathematics Study (Westbury, Russell and Travers, 1989). For example, in the US, the number of female citizens who earned a doctoral degree in mathematics in the year 1986 to 1987 was only 20 per cent (National Center for Education Statistics, 1992). More recently it has been noted in the US that gender differences in participation are becoming negligible in all parts of the curriculum except calculus where equal representation remains a problem (Streitmatter, 1994). In the UK, while the proportion of female students taking mathematics courses is increasing, it still remains disproportionately lower than that of males (Shuard, 1983). In the majority of the countries of the Second International Mathematics Study, there are two or three times as many boys as girls in advanced mathematics courses (Westbury, Russell and Travers, 1989).

Developing Countries

Cultural norms also contribute to a deficit in female enrolment in institutions of learning in many developing countries. Families which can afford to send only one child to school will invariably send a boy, since he is considered the future breadwinner for the family. According to Anbesu and Junge (1988), a survey of the Gojjam region of Ethiopia showed that only 10 per cent of all children between the ages of 7 and 16 years were enrolled in school and, of these pupils, two-thirds were boys. Similarly, Guy (1992) found that only 39 per cent of Papua New Guinean secondary-school students were female. The literature from many other developing countries confirms this trend (Grace, 1991). More recently the Population Action International study reported that some eighty-five million more boys than girls receive elementary and secondary education throughout the world (*Chicago Tribune*, 1994). These disparities become even more acute at the tertiary level (Kaeley, 1990). For example, citing a number of studies from Papua New Guinea, Kaeley writes that female students in the University of Papua New Guinea in the years 1966, 1977, and 1988 comprised only 10 per cent, 15 per cent, and 18 per cent, respectively, of the total university population.

Mathematics Performance

Developed Countries

The First International Study of Achievement in Mathematics, which was conducted in more than twelve developed countries, showed mixed results concerning gender differences in mathematics performance (Thorndike, 1967). The achievement of boys was better than that of girls at age 13, particularly on verbal, as opposed to computational, problems. However, there was substantial variation between countries, with gender differences being largest in Belgium and Japan, and least in the US and Sweden (Thorndike, 1967). Thorndike attributes the small differences in the latter countries to the improved role of women in both societies compared with Japan and Belgium.

To concentrate for the moment on the US, gender differences in mathematics performance have declined over the years since 1967 to the point where their magnitude is now very small (Hyde *et al.*, 1990). In a meta-analysis of studies conducted between 1960 and 1983, Feingold (1989) observed that gender differences in cognitive performance are disappearing, although the disparity at the upper levels of performance in high-school mathematics has remained constant throughout the same period. These results were corroborated by Friedman (1989) in her meta-analysis of studies conducted between 1974 and mid-1987. Furthermore, she predicts that gender differences in mathematics performance at all levels of learning will vanish altogether in forty years.

From data compiled in the Second International Mathematics Study, Hanna, Kündiger and Larouche (1990) compared sex differences in mathematics performance of students in the final year of secondary school in fifteen north American, European and east Asian countries. They showed that differences were negligible for those countries where there were high levels of home support for both sexes, whereas countries where substantial sex differences were observed had lower levels of home support. They suggested that since society has traditionally reinforced boys' participation in mathematics, the degree of support provided by the home may be a more crucial indicator of success in mathematics for girls than it is for boys (Hanna, Kündiger and Larouche, 1990). This could mean that, when societal norms are counter-balanced by family expectations, gender differences disappear. Furthermore, the finding, that gender differences in mathematics performance vary between countries, runs counter to traditional arguments that attribute boys' superior mathematical performance to their biology, and suggests instead that cultural factors offer a more viable explanation (Hanna, Kündiger and Larouche, 1990).

Developing Countries and Ethnic Minorities in Developed Countries

The cultures of a number of ethnic minorities in the developed world are either female-dominated or gender-equitable (Driver, 1980). Even in cases where the indigenous culture is male-dominated, pressures by the dominant culture are producing change. As the examples below illustrate, in some ethnic groups, this results in female performance in mathematics which is as good as, or better than, their male counterparts.

From a meta-analysis of 100 studies of gender differences in mathematics performance in the US, Hyde *et al.* (1990) demonstrated that there were no gender differences in the in-class mathematics performance of African-American and Hispanic-American students. A slight difference for Asian-Americans was discovered and it favoured female students. Only for white students was there any evidence of superior male performance, although the gap was small. In another study, this time of Hawaiian students between the ages of 9 and 16 years, Brandon, Newton, and Hammond (1987) concluded that girls outperformed boys in mathematics in all the observed ethnic groups and that the female advantage was significantly larger in the Philippine, Hawaiian, and Japanese samples than it was in the Caucasian sample (and this has continued, see also Brandon, Jordan and Higa, Chapter 11, this volume).

Another example of the influence of culture on mathematics performance is provided by Geoffrey Driver (1980). He studied 2300 secondary-school graduates of both sexes, including white students and students of West Indian descent, in five multiracial secondary schools in the UK. He observed that West Indian girls outperformed West Indian boys in English language, mathematics, and science subjects. White boys excelled over White girls, but not as markedly. Out of all four groups, the highest mean results were obtained by West Indian girls, with West Indian boys doing less well than White boys and performing at the level of the White girls. Driver argues that the superior performance of West Indian girls is probably due to their culture. Citing a number of studies, he suggests that in rural Jamaica, the women rather than the men assume responsibility for the family's survival. Upon immigration, West Indian families retain this custom and this is reflected in women's superior academic performance.

India is a developing country in which male dominance is the societal norm. In a survey of mathematics attainment in Indian schools, Kulkarni, Naidu, and Arya (1969) observed inferior achievement by girls relative to boys in most states. Notwithstanding, there were some remarkable exceptions, for example, the highly-developed Mangalore region of Mysore State where females outperformed males.

The influence of culture on the role of women in society, on the cognitive development of children, and on the learning of mathematics can also be found in Papua New Guinea and some South Pacific Island nations. In these countries, there are a number of matrilineal communities (see, for example, Melbourne University Press and University of Papua New Guinea, 1972) in which women have more power and authority in the family than the men. Under such circumstances, girls are treated with respect in the classroom as well as in the society as a whole. And indeed we find that females exhibit equal, if not better, performance in mathematics compared with males (Kaeley, 1990).

In 1983, Lancy conducted a major research project on the cognitive development of primary-school children in Papua New Guinea. He observed that even in patrilineal communities, there are no sex differences in cognitive development and the learning of mathematics. However, an analysis of the nationwide grade 6 (11–12 years of age) mathematics examination results does not support his conclusions, since boys' performance was found to be significantly better than that of girls (Measurement Services Unit, 1985). A study by Clarkson (1983), which focused on gender disparities in the learning of mathematics by 15-year-olds in Papua New Guinea, also showed no differences in the performance of boys and girls. An earlier study of a situation in which 15-year-old boys excelled over girls in mathematics noted that, given the right environment, achievement levels are equal (Silvey, 1978). At the tertiary level, Kaeley (1988) has observed that there were no gender differences in post-secondary mathematics performance at the University of Papua New Guinea. The reasons he suggested for this apparent discrepancy with the results of other studies are methodological: first, included in the sample were a number of female students from the matrilineal communities; second, as proportionally fewer females than males continue on to secondary and post-secondary education, they are not representative of the female population of the country as a whole; and third, half the students in the sample were engaged in distance education and thus were free from teacher-effect. Furthermore, he speculated that, since formal mathematics has only recently been introduced in Papua New Guinea, it was foreign to all students and therefore impervious to gender. Interestingly though, the only indigenous mathematics doctoral degree was earned by a woman.

Conclusion

Overall, it is evident that enrolment differences in developing countries predominantly favour male students. Even in the majority of developed countries, cultural influences encourage male students to take additional mathematics courses beyond the stage at which they are compulsory. This situation, however, is slowly changing and gender differences in mathematics performance are decreasing as women's role in society improves. There are certain developing countries, as well as a number of ethnic communities within developed societies, in which women have more power and authority than men. In these societies, either there are no apparent gender differences, or female students outperform their male counterparts in mathematics. The main conclusion of this chapter is that culture is one of the most important factors contributing to gender differences in the learning of mathematics.

References

ANBESU, B. and JUNGE, B. (1988) 'Problems in Primary School Participation and Performance in Bahir Dar Awraja', (mimeograph), Addis Ababa, UNICEF and Ministry of Education.

BRANDON, P.R., NEWTON, B.J. and HAMMOND, O.W. (1987) 'Children's mathematics achievement in Hawai'i: Sex differences favouring girls', *American Educational Research Journal*, 24, pp. 437–61.

Chicago Tribune (1994) 'Study finds male-female education gap', in *Chicago Tribune*, January 31, Chicago, Tribune Publishing Company, p. 7.

CLARKSON, P. (1983) 'Papua New Guinea students' causal attributions for success and failure in mathematics', Report No. 27, Lae, Mathematics Education Centre, Papua New Guinea University of Technology.

CONNORS, E.A. (1985) 'Report on the 1985 survey of new doctorates', *Notices of the American Mathematical Society*, 32, 6, pp. 768–71.

DRIVER, G. (1980) 'How West Indians do better at school (especially the girls)', in *New Society*, 51, 902, pp. 111–14.

FEINGOLD, A. (1989) 'Cognitive gender differences are disappearing', *American Psychologist*, 43, 2, pp. 95–103.

FENNEMA, E. (Ed) (1985) 'Explaining sex-related differences in mathematics: Theoretical models', *Educational Studies in Mathematics*, 16, 3, pp. 303–20.

FRIEDMAN, L. (1989) 'Mathematics and the gender gap: A meta-analysis of recent studies on sex differences in mathematical tasks', *Review of Educational Research*, 59, 2, pp. 185–213.

GRACE, M. (1991) 'Gender issues in distance education', in EVANS, T.D. and KING, B. (Eds) *Beyond the Contemporary Writing in Distance Education*, Geelong, Deakin University Press.

GUY, R.K. (1992) 'Privileging others and otherness in research in distance education', in EVANS, T.D. and PARER, J. (Eds) *Research in Distance Education Two*, Geelong, Deakin University Press.

HANNA, G., KÜNDIGER, E. and LAROUCHE, C. (1990) 'Mathematical achievement of Grade 12 girls in fifteen countries', in BURTON, L. (Ed) *Gender and Mathematics: An International Perspective*, London, Cassell, pp. 87–97.

HYDE, J.S., FENNEMA, E. and LAMON, S.J. (1990) 'Gender differences in mathematics performance: A meta-analysis', *Psychological Bulletin*, 107, 2, pp. 139–55.

KAELEY, G.S. (1988) 'Sex differences in the learning of post-secondary mathematics in a neo-literate society', *Educational Studies in Mathematics*, 19, pp. 435–57.

KAELEY, G.S. (1990) Distance Versus Face-to-Face Learning: A Mathematics Test Case,

Unpublished PhD Thesis, Waigani, University of Papua New Guinea.

KIMBALL, M.M. (1989) 'A new perspective on women's math achievement', *Psychological Bulletin*, 105, 2, pp. 198–214.

KULKARNI, S.S., NAIDU, C.A.S. and ARYA, M.L. (1969) 'A survey of mathematics achievement in Indian schools', *Indian Educational Review*, 4, 1, pp. 1–33.

LANCY, D.E. (1983) *Cross-cultural Studies in Cognition and Mathematics*, New York, Academic Press.

LEDER, G. (1980) 'Bright girls, mathematics and fear of success', *Educational Studies in Mathematics*, 11, 4, pp. 411–22.

LEDER, G. (1985) 'Sex-related differences in mathematics: An overview', in FENNEMA, E. (Ed) 'Explaining sex related differences in mathematics: Theoretical models', *Educational Studies in Mathematics*, 16, 3, pp. 304–9.

LUMMIS, M. and STEVENSON, H.W. (1990) 'Gender differences in beliefs and achievement: A cross-cultural study', *Developmental Psychology*, 26, pp. 254–63.

MEASUREMENT SERVICES UNIT (1985) *Analysis of 1982 Grade 6 Examination Results*, Waigani, Department of Education.

MEECE, J.L., PARSON, J.E., KACZALA, C.M., GOFF, S.B. and FUTTERMAN, R. (1982) 'Sex differences in mathematical achievement: Toward a model of academic choice', *Psychological Bulletin*, 91, 2, pp. 324–48.

MELBOURNE UNIVERSITY PRESS AND UNIVERSITY OF PAPUA NEW GUINEA (1972) *Encyclopedia of Papua New Guinea*, 2.

NATIONAL CENTER FOR EDUCATION STATISTICS (1992) *Race/Ethnicity Trends in Degrees Conferred by Institutions of Higher Education: 1980–81 Through 1989–90*, Washington, DC, US Government Printing Office.

SHUARD, H.B. (1982) 'Differences in mathematical performance between boys and girls', in COCKCROFT, W.H. (Chairman) *Mathematics Counts*, London, Her Majesty's Stationery Office.

SILVEY, J. (1978) *Academic Success in PNG High Schools*, ERU Research Report No. 26, Waigani, University of Papua New Guinea.

STREITMATTER, J. (1994) *Toward Gender Equity in the Classroom: Everyday Teachers' Beliefs and Practices*, Albany, State University of New York Press.

THORNDIKE, R.L. (1967) 'Mathematics test and attitude inventory scores', in HUSÉN, T. (Ed) *International Studies of Achievement in Mathematics II*, Stockholm, Almqvist and Wiksell, pp. 21–48.

WESTBURY, I., RUSSELL, H.H. and TRAVERS, K.J. (1989) 'The content of the implemented mathematics curriculum', in TRAVERS, K.J. and WESTBURY, I. (Eds) *The IEA Study of Mathematics I: Analysis of Mathematics Curricula*, Oxford, Pergamon Press, pp. 167–202.

XU, J. and FARRELL, E. (1992) 'Mathematics performance of Shanghai high school students: A preliminary look at gender differences in another culture', *School Science and Mathematics*, 92, pp. 442–5.

Chapter 11

Why Hawai'i Girls Outperform Hawai'i Boys: The Influence of Immigration and Peer Culture

Paul Brandon, Cathie Jordan and Terry Ann Higa

Gender differences in mathematics achievement in the United States historically have favoured boys. In the State of Hawai'i, however, these differences have favoured girls. What is different about Hawai'i that might help account for the state's unique pattern of gender differences in mathematics achievement?

In this chapter, we suggest that some characteristics of Hawai'i's immigrant and native Hawaiian populations may have enhanced girls' educational achievement and negatively affected boys' achievement. First, we give a brief summary of the findings on gender differences in mathematics achievement across the United States and in Hawai'i and discuss gender differences in verbal skills among children nationwide and in Hawai'i. Second, we summarize research on the effects of immigration on males and females and present the thesis that, in Hawai'i, such effects may have manifested themselves in Hawai'i's gender differences in educational achievement. Third, we summarize anthropological theory on school compliance and ethnographic data on peer culture among native Hawaiian youngsters, and we suggest how these are applicable to gender differences among native Hawaiians.

Gender Differences

Maccoby and Jacklin, in their landmark survey of research on gender differences in mathematics achievement, concluded that American boys and girls were 'similar in their early acquisition of quantitative concepts, and their mastery of arithmetic during the grade-school years' (Maccoby and Jacklin, 1974, p. 352), but that boys' mathematical skills increased faster than girls' after about 12 years of age. Recent meta-analyses have shown that differences in mathematics achievement have decreased and largely been eliminated (Friedman, 1989; Hyde, Fennema, and Lamon, 1990) except among high-school and post-secondary students on some mathematics subtests, on which boys have continued to show somewhat higher levels of achievement than girls.

In Hawai'i, studies have consistently shown elementary- and secondary-school girls outperforming boys on mathematics tests (Brenner, 1984; Holmes, 1968; Kamehameha Schools/Bishop Estate, 1983; Marshall, 1927; Stewart, Dole and Harris, 1967). Brandon, Newton and Hammond (1987) reported an analysis of 1982 and 1983 Stanford Achievement Test results from Hawai'i's statewide testing programme. For children of Filipino, native Hawaiian, Japanese, and white ancestry, results on three mathematics subtests for Grades 4, 6, and 8 (ages 9–14 years) and one subtest for Grade 10 (ages 15–16 years) showed that girls outperformed boys in seventy-six of eighty comparisons. Results across ethnic groups, subtests, and testing years were combined and effect sizes (that is, the girls' mean minus the boys' mean divided by the standard deviation of both groups combined) were calculated. The effect sizes favoured girls and were greater among older children than among younger children, with sizes ranging from 0.12 for Grade 4 to 0.26 for Grade 10. A finding that is particularly relevant for this chapter is that differences favouring girls were greater among children of non-White ancestry than among White children.

Brandon and Jordan (1994) have expanded on these findings in two ways. First they analysed 1991 Hawai'i achievement test results, showing that gender differences in mathematics achievement among Hawai'i school children have continued. Second they compared findings for Hawai'i on the 1990 National Assessment of Educational Progress state-level eighth-grade mathematics assessment with the findings of thirty-six other states, the District of Columbia, and two territories. Hawai'i was the only one of the forty participating jurisdictions in which girls' total-test mean scores were significantly higher than those of the boys. Therefore, Brandon and Jordan concluded that the Hawai'i gender differences are unique among the American states.

Another area in which boys' and girls' achievement has traditionally and widely been compared is verbal skills. In a meta-analysis of studies on differences in verbal skills, Hyde and Linn (1988) found an effect size of 0.11. They concluded that 'overall, studies of gender differences in verbal ability support the conclusion that these differences are now negligible' (Linn and Hyde, 1989, p. 18). In contrast, analyses conducted for this chapter show that Hawai'i Stanford Achievement Test statewide testing results for four grades in reading in 1991 show effect sizes ranging from 0.12 to 0.24, with a mean effect size of 0.18. Thus, the results for Hawai'i suggest larger differences favouring girls than the national results, and, together with the results found on gender differences in mathematics achievement, suggest a general educational achievement advantage for Hawai'i girls relative to boys.

What explanations for Hawai'i educational gender differences might we glean from the literature? In the next section, we suggest that immigration might have affected Hawai'i boys' and girls' educational achievement and attainment levels.

The Impact of Immigration

'Legal immigration to the United States . . . has been dominated by females for the past half century' (Pedraza, 1991, p. 304), yet few studies have focused on women immigrants (Pedraza-Bailey, 1990). Nevertheless, by examining reports of studies on immigrants published in the psychological, sociological, and anthropological literature, we have found sufficient literature to support the claim that a theme has emerged: immigration into western nations may be more beneficial for females than for males, at least in the short run, with the consequence that female immigrants experience greater educational success than male immigrants.

Female Immigrants May Adapt More Easily than Males

Studies of Jewish, Irish, and Caribbean immigrants to the US and England have suggested that female immigrants experience less difficulty than males in adapting to their new lives. Summarizing the status of Jewish working-class male immigrants in the late-nineteenth and early-twentieth centuries, Hertzberg said the 'evidence of the destruction of the role of the father is overwhelming' (Hertzberg, 1989, p. 196). Less than 20 per cent of the men during this period earned sufficient income to support their families, and many deserted their families, resulting in the development (or, in Hertzberg's words, the 'invention') of the 'Jewish mother' as the family's protector (Hertzberg, 1989, p. 196). Among Irish immigrants, men experienced a similar 'decline in status and power within their families as a result of migration, pushing women — wives and mothers — into authoritative roles far greater than they had experienced' in Ireland (Diner, 1983, p. 46). Migration was a '"liberating" experience for Irish women. They consciously had chosen to leave a society that offered them little as women to embrace one that proffered to them greater opportunity . . .' (Diner, 1983, p. 140). Pessar (1986) found that women immigrants to the US from the Dominican Republic were reluctant to return to the Dominican Republic because of the gains they had experienced after migrating. As Pessar said about these women immigrants, 'in contrast to men, migration does not rupture the social sphere in which women are self-actualized' (Pessar, 1986, p. 276). According to Foner (1978), among Jamaican immigrants to England, immigration was more positive for women than men because the women had become less dependent upon men.

Studies of Chinese immigrants have shown similar post-immigration male–female differences. In a study of Chinese Americans, Yu (1984) reported that the relationship between life satisfaction and acculturation was stronger for Chinese-American females than males. Huang compared male and female Chinese-American students and reported that females 'prove to be more sensitive to Westernization and more balanced in their acculturation process' (Huang, 1956, p. 25) than males. In an anthropological study of first-generation Chinese Americans' dating behaviour, Weiss (1970) reported that females had adjusted more quickly and easily than males to American social customs, that they were more accepted by Whites as dating partners and possible mates, and that they often thought that Chinese-American males were inept dating partners.

In studies of Asian immigrants living in Hawai'i, similar patterns suggesting less difficult adjustment among females than males have been found. Arkoff, Meredith and Iwahara (1962) concluded that male Japanese immigrants might have acculturated less easily or quickly to American culture than female Japanese immigrants. Among later generations of Japanese Americans in Hawai'i, findings on gender differences suggest that the effects of differential adaptation have persisted. For example, Meredith and Meredith (1973) compared Japanese Americans and White Americans and showed that Japanese-American males in their study were more introverted than the White males but that no such differences were found between the Japanese-American and White females. Similarly, Bartos and Kalish (1961) showed that among college-campus leaders, the proportion of Japanese-American females was almost three times the proportion of Japanese-American males.

Immigration May Affect Educational Achievement and Attainment Levels

Published evidence for gender differences in immigrants' educational achievement is sparse and presents a mixed picture. A summary of late-nineteenth and early-twentieth century immigrants to the US (Olneck and Lazerson, 1974) showed that immigrant girls in most ethnic groups had completed more years of high-school than immigrant boys, but among Russian Jews and southern Italians, boys had completed more years of school than girls. Brandon (1991) analysed data from the 1986 follow-up of the national 'High School and Beyond' survey. He concluded that Asian-American females have reached high levels of educational attainment more quickly than Asian-American males and that these differences have been largest among Chinese Americans, Filipino Americans, and recent immigrants or their children. Duran and Weffer, however, examined educational data on Mexican-American immigrants and found that 'girls had lower achievement before high school and the differential increased through the high school years' (Duran and Weffer, 1992, p. 177).

Such findings suggest that group-specific cultural attitudes toward the education of the sexes may have influenced immigrants' educational achievement or attainment levels. Among late-nineteenth and early-twentieth century immigrant Russian Jews, for example, the education of males was highly valued, and among immigrant southern Italians during the same period, the education of females was disparaged. Campbell and Connolly (1987) found that parents' expectations of White American students' secondary and post-secondary performance were greater for boys than for girls. Among Asian Americans, however, Campbell and Connolly found that parents' expectations were the same for both sexes. Among pre-colonial Filipinos, men and women were regarded in an egalitarian manner, and today the education of Filipino women is highly valued (Alcantara, 1990). The effects of these attitudes toward Filipino women's education are manifested in the high educational attainment rates among Filipino women in the Philippines, where 'the literacy rate among women is equal to that of men' (Alcantara, 1990). They are also manifested in the educational attainment rates of young Filipino immigrant women. In Table 11.1, we show analyses of data taken from the US Census Bureau's computer runs of unpublished 1980 census results (US Bureau of the Census, 1986) on the educational attainment of male and

Table 11.1: Percentages of native-born and immigrant Asian-American males and females aged 20–9 years in 1980 who had graduated from college

	Native-born		Immigrants	
	Male	**Female**	**Male**	**Female**
Filipinos	10.8	14.8	23.1	35.3
Chinese	40.3	40.5	36.8	31.6
Vietnamese	15.7	20.0	9.8	6.0
Japanese	31.2	34.5	31.5	23.7
Koreans	23.5	21.9	23.7	15.0
Asian Indians	28.2	22.5	53.8	43.3

Source: US Bureau of the Census (1986)

female immigrant and native-born Asian Americans aged 20–9 years in six ethnic groups. As can be seen, gender differences favouring female immigrants are found only among Filipinos.

Immigration and Educational Achievement

Hawai'i is the youngest American state, and much of its population is comprised of the most recent generations of immigrants to the US. Before the mid-1800s, when many European immigrants had already entered the US mainland, only small numbers of non-native people had come to Hawai'i to stay. From 1852 to 1930, however, about 400 000 immigrants were brought to the state to work on plantations (Fuchs, 1961). In recent years, Hawai'i has continued as a home for new immigrants; for example, Hawai'i's 1987 population comprised only 0.44 per cent of the total US population, but 1.13 per cent of all immigrants to the US in 1987 chose Hawai'i as their intended residence (National Technical Information Service, 1990). Most recently, over half of these immigrants have come from the Philippines (Caces, 1985; Department of Business and Economic Development, 1988).

If immigration affects females more positively than males, the relatively high proportions of immigrants in Hawai'i throughout its recent history might account at least in part for its unique gender differences favouring girls in mathematics achievement. The argument, supported by the findings and studies summarized in this chapter so far, goes as follows. First, gender differences in Hawai'i are found in both mathematics and verbal skills, suggesting a general educational achievement advantage for girls. Second, these differences are greater among non-Whites than among Whites, leading us to focus in Hawai'i on Asian immigrants, the predominant non-White immigrant group. Third, among several national or ethnic groups nationwide (including Asians), immigration may have been more beneficial for women than for men, at least in the short run. Fourth, the positive effect of immigration on females in Hawai'i has been maintained beyond the first generation and has been seen in immigrants' descendants. Finally, Filipinos, who comprise Hawai'i's largest recent immigrant group, have high regard for the education of females, thus reinforcing long-standing Hawai'i trends favouring females' outperformance of boys in school. This evidence forms the foundation for our suggestion that the effects of immigration help account for gender differences in educational achievement favouring girls in Hawai'i.

School Compliance and Native Hawaiian Peer Culture

Rationales for School Compliance

Anthropologists and sociologists have pointed out that schools require children to give up independence and expect them to acknowledge the authority of strangers who have no pre-established legitimacy. All children have to decide whether they are going to comply with what schools and teachers ask of them. Especially for minority culture children, much of what the teacher and school require may seem strange and even sometimes offensive. Since some minimal degree of student compliance is a necessary condition for success in school, the decision about whether to comply with school rules and teacher requirements is a crucial one.

John D'Amato (1986, 1993) has developed a useful perspective on school compliance, based mainly on his work with primary-grade native Hawaiian children. According to D'Amato, there are two kinds of rationales for complying with the teacher's wishes, 'structural' and 'situational'. A structural rationale for compliance involves knowledge of, and belief in, external advantages of complying that are of such magnitude that children are willing to tolerate even very punishing conditions internal to the school context in service of external goals. A structural rationale for compliance could involve, for example, the conviction that success in school leads to success in life, or parental and familial pressure on the student to comply with school requirements and excel in school, even at the sacrifice of other culturally important roles and values.

A situational rationale for compliance, on the other hand, involves factors intrinsic to the school or classroom. Children who do not have a strong structural rationale for complying with school demands may nevertheless comply if doing so brings them valued payoffs within the school setting and does not violate their sense of identity or their beliefs about appropriate behaviour. In particular, at least with native Hawaiian children (D'Amato, 1993), when classroom practice and the teacher's claim to authority do not conflict with the structure and demands of the children's peer groups, then even children who have little in the way of structural rationales for compliance may cooperate with the teacher.

Peer Groups and Rationales for School Compliance

How does this relate to male–female differences in the educational achievement scores of native Hawaiian students? We suggest that such gender differences may occur because of differences in the availability of structural and situational rationales for compliance and achievement.

Hawaiian families and kinship networks traditionally were generationally organized; many still tend toward this kind of organization (D'Amato, 1986; Howard, 1974). One consequence of this organization is that adults interact mostly with adults, and children with other children. After infancy, children are socialized to become part of an interdependent, interacting group of siblings which bears collective responsibility for much of the work of the household, including considerable childcare (Gallimore, Boggs, and Jordan, 1974; Jordan, 1981). This sibling group is central to the lives of many Hawaiian families. In the sibling group, an older child, usually the oldest girl, often acts as helper or 'right hand' to her mother (D'Amato, 1986; Gallimore, Boggs and Jordan, 1974). She is responsible in a special way for looking out for her younger siblings and for seeing that the work of the sibling group gets done. She has a special place in the affections of her siblings, as well as a special degree of influence with them. In addition, because of her relationship with her mother, she is a link between the generations.

Starting in later childhood, and becoming increasingly important in the teenage years, companion groups of close-age, same-sex peers become important, especially for boys (Gallimore, Boggs and Jordan, 1974). The peer companion group tends to remain central for males into early adulthood, even after family formation and marriage. It exerts a powerful 'pull' for boys and young men which is only gradually overcome by the needs and ties of the family and kinship network. In these same years, females are becoming increasingly involved in kin and family concerns. Kinship networks tend to be held together by ties between females, with males being more peripheral (D'Amato, 1986). Because of the importance of links among female kin, young women may be more involved in cross-generational interactions than young men are.

In school, there is evidence of a strong peer orientation among both boys and girls (Gallimore, Boggs and Jordan, 1974). In D'Amato's (1986) study of a class of low-income, elementary-school native Hawaiian children, girls and boys had separate, differently organized peer groups with different kinds of objectives. Both boys' and girls' peer groups operated in a dynamic of rivalry in which individuals strove to establish and maintain themselves in positions of equality or near equality with other members of their peer group. However, the boys in the study engaged in rivalry for dominance among themselves, while the girls engaged in rivalry for central social positions within their peer group. Also, for girls, closeness to the teacher seemed to be an important counter in their game of centrality, in a way that did not have a parallel in the boys' dominance game. In addition, in some classrooms, one girl appears to take on the position of chief helper to the teacher, echoing the role of mother's helper. In D'Amato's study, this was the same girl who occupied the most central position in the peer group. This girl thus crossed the generational line, just as the mother's helper does in the family, and may have provided a link between the teacher and the girls' peer group that did not exist for boys.

These patterns in children's lives at home and at school may, at least in part, be responsible for gender differences in achievement among native Hawaiian youngsters. For Hawaiians, as a group, there is little structural rationale for compliance and achievement in school. With a history of conquest and colonization by the United States, they are at the lower end of the economic spectrum. They have suffered long-term discrimination against their culture and language by the dominant group. This kind of experience does not inspire much trust in schools as an institution of the dominant society, or faith in schooling as a realistic road to success in life.

That being the case, we need to look to the classroom setting and the dynamics of interaction in school to determine the relative strength of situational rationales for (or against) school compliance. Since peer orientation among native Hawaiian children is strong and the peer group important, its influence must be considered carefully. We suggest that the way girls' peer groups operate is more compatible with the ordinary school situation than is the organization and operation of boys' peer groups. It thus may constitute a kind of situational rationale for compliance that does not exist to the same degree for boys. Among younger children, the dominance rivalries and struggles of the boys may present more serious and disruptive challenges to teachers and more difficult barriers to compliance than does the centrality-oriented rivalry of girls. This is especially true since girls can actually incorporate closeness to the teacher into their centrality game, and since the central girl in the peer group can also serve as a facilitator and bridge to friendly relations with the teacher. Among older students, it is possible that, as girls' concerns and energies turn more toward family obligations and kin ties, the power of their school-based peer groups may diminish somewhat, but that it remains as strong or stronger for older boys as it is for younger boys. The values and interactions of the peer group may have much more appeal for boys than anything the traditionally organized school usually offers in the way of situational rationales for compliance.

Peer Group Culture and Other Ethnic Groups

The importance of the male peer group and its rivalry games seems to have precedents in native Hawaiian culture and to appear most dramatically in schools with high native Hawaiian populations, but to a certain extent, it is also a phenomenon of Hawai'i's 'local' youth culture as a whole. Thus, it may affect all of Hawai'i's male students to a greater or lesser degree, and we can speculate about how this fact may relate to the relative sizes of gender differences in achievement scores for other ethnic groups.

Filipino children are more likely than Japanese or Caucasian children to live in the same neighbourhoods and attend the same schools as native Hawaiian youngsters. Therefore, factors similar to those associated with the male peer-group organization for Hawaiians may have more effect on Filipino boys and less effect on Japanese and Caucasian boys, resulting in a larger differential between Filipino boys' and girls' achievement.

Achievement-test effect sizes favouring girls may be the smallest for Caucasian children because many are immigrants from the mainland or, as the children of military families, are in Hawai'i for only relatively short periods of time. Thus, Caucasian students are perhaps the least likely to participate in the local school culture, including the peer-group organization of boys with its attendant potential for impacting the degree of school compliance.

Conclusion

In this chapter, we have summarized literature suggesting that differential adaptation to their new homeland by male and female immigrants to the state of Hawai'i and differential school compliance among native Hawaiian boys and girls may help account for gender differences favouring Hawai'i girls on both mathematics and verbal-achievement tests. The effects of immigration were seen in the state as early as the 1920s and, in recent years, may have continued because of the high proportion of Filipino immigrants to the state. We speculate that differential school compliance among native Hawaiian boys and girls may also help account for gender differences in achievement among boys and girls of the other ethnic groups in the state. Of course, in addition to immigration and school compliance, there may very well be other sociological, anthropological, psychological, or historical factors which also affect gender differences in Hawai'i. These intriguing differences in achievement, unique among the American states, deserve further examination by educational researchers, practitioners, and policy makers.

References

ALCANTARA, A.N. (1990) Gender Differentiation: Public versus Private Power in Family Decision Making in the Philippines, Unpublished doctoral dissertation, University of Hawai'i, Honolulu.

ARKOFF, A., MEREDITH, G. and IWAHARA, S. (1962) 'Dominance-deference patterning in motherland-Japanese, Japanese-American, and Caucasian-American students', *Journal of Social Psychology*, 52, pp. 368–420.

BARTOS, O.J. and KALISH, R.A. (1961) 'Sociological correlates of student leadership in Hawai'i', *Journal of Educational Sociology*, 35, pp. 65–72.

BRANDON, P.R. (1991) 'Gender differences in young Asian Americans' educational attainment', *Sex Roles*, 25, pp. 47–63.

BRANDON, P.R. and JORDAN, C. (1994) 'Gender differences favouring Hawai'i girls in mathematics achievement: Recent findings and hypotheses', *Zentralblatt für Didaktik der Mathematik*, 26, 1, pp. 18–21.

BRANDON, P.R., NEWTON, B.J. and HAMMOND, O.W. (1987) 'Children's mathematics achievement in Hawai'i: Sex differences favouring girls', *American Educational Research Journal*, 24, pp. 437–61.

BRENNER, M.E. (1984) Standardized Arithmetic Testing at KEEP (1975, 1977), Unpublished manuscript, Kamehameha Schools/Bishop Estate, Center for Development of Early Education, Honolulu.

CACES, M.F. (1985) Personal Networks and the Material Adaptation of Recent Immigrants:

A Study of Filipinos in Hawai'i, Unpublished doctoral dissertation, University of Hawai'i, Honolulu.

CAMPBELL, J.R. and CONNOLLY, C. (1987) 'Deciphering the effects of socialization', *Journal of Educational Equity and Leadership*, 7, pp. 208–22.

D'AMATO, J. (1986) 'We Cool, Tha's Why.': A Study of Personhood and Place in a Class of Hawaiian Second Graders, Unpublished doctoral dissertation, University of Hawai'i, Honolulu.

D'AMATO, J. (1993) 'Resistance and compliance in minority classrooms', in JACOB, E. and JORDAN, C. (Eds) *Minority Education: Anthropological Perspectives*, Norwood, NJ, Ablex, pp. 104–10.

DEPARTMENT OF BUSINESS AND ECONOMIC DEVELOPMENT (1988) *The State of Hawai'i Data Book*, Honolulu.

DINER, H.R. (1983) *Erin's Daughters in America: Irish Immigrant Women in the Nineteenth Century*, Baltimore, John Hopkins University Press.

DURAN, B.J. and WEFFER, R.E. (1992) 'Immigrants' aspirations, high school process, and academic outcomes', *American Educational Research Journal*, 29, 1, pp. 163–81.

FONER, N. (1978) *Jamaica Farewell: Jamaican Migrants in London*, Berkeley, University of California Press.

FRIEDMAN, L. (1989) 'Mathematics and the gender gap: A meta-analysis of recent studies on sex differences in mathematical tasks', *Review of Educational Research*, 59, pp. 185–213.

FUCHS, L.H. (1961) *Hawaii Pono: A Social History*, New York, Harcourt and Brace.

GALLIMORE, R., BOGGS, J. and JORDAN, C. (1974) *Culture, Behavior and Education: A Study of Hawaiian-Americans*, Beverly Hills, CA, Sage.

HERTZBERG, A. (1989) *The Jews in America*, New York, Simon and Schuster.

HOLMES, G.C. (1968) A Study of Value and Attitudinal Correlates of School Achievement and Success in Nnakuli, Unpublished doctoral dissertation, Syracuse University, Syracuse, NY.

HOWARD, A. (1974) *Ain't No Big Thing: Coping Strategies in a Hawaiian-American Community*, Honolulu, University Press of Hawai'i.

HUANG, L. (1956) 'Dating and courtship innovations of Chinese students in America', *Marriage and Family Living*, 18, pp. 25–9.

HYDE, J.S., FENNEMA, E. and LAMON, S.J. (1990) 'Gender differences in mathematics performance: A meta-analysis', *Psychological Bulletin*, 107, pp. 139–55.

HYDE, J.S. and LINN, M.C. (1988) 'Gender differences in verbal ability: A meta-analysis', *Psychological Bulletin*, 104, pp. 53–69.

JORDAN, C. (1981) Educationally Effective Ethnology: A Study of the Contributions of Cultural Knowledge to Effective Education for Minority Children, Occasional paper of the Center for Development of Early Education, Honolulu, Kamehameha Schools/Bishop Estate.

KAMEHAMEHA SCHOOLS/BISHOP ESTATE (1983) *Native Hawaiian Educational Assessment Project Final Report*, Honolulu.

LINN, M.C. and HYDE, J.S. (1989) 'Gender, mathematics, and science', *Educational Researcher*, 18, 8, pp. 17–19, 22–7.

MACCOBY, E.E. and JACKLIN, C.N. (1974) *The Psychology of Sex Differences*, Stanford, Stanford University Press.

MARSHALL, E.L. (1927) A Study of the Achievement of Chinese and Japanese Children in the Public Schools of Honolulu, Unpublished master's thesis, University of Hawai'i, Honolulu.

MEREDITH, G.M. and MEREDITH, C.G.W. (1973) 'Acculturation and personality among Japanese-American college students in Hawai'i', in SUE, S.S. and WAGNER, N.N. (Eds) *Asian-Americans: Psychological Perspectives*, Palo Alto, CA, Science and Behavior Books, pp. 104–10.

NATIONAL TECHNICAL INFORMATION SERVICE (1990) *1989 Statistical Yearbook of the Immigration and Naturalization Service*, Washington, DC, US Department of Commerce.

OLNECK, M.R. and LAZERSON, M. (1974) 'The school achievement of immigrant children: 1900–1930', *History of Education Quarterly*, 14, pp. 453–82.

PEDRAZA, S. (1991) 'Women and migration: The social consequences of gender', *Annual Review of Sociology*, 17, pp. 303–25.

PEDRAZA-BAILEY, S. (1990) 'Immigration research: A conceptual map', *Social Science History*, 14, 1, pp. 43–67.

PESSAR, P.R. (1986) 'The role of gender in Dominican settlement in the United States', in NASH, J. and SAFA, H. (Eds) *Women and Change in Latin America*, South Hadley, MA, Bergin and Garvey, pp. 273–93.

STEWART, L.H., DOLE, A.A. and HARRIS, Y.Y. (1967) 'Cultural differences in abilities during high school', *American Educational Research Journal*, 4, pp. 19–29.

US BUREAU OF THE CENSUS (1986) 'Numbers of years of school completed, by immigrant status, age group, and ethnic group', Unpublished computer-generated tables.

WEISS, M. (1970) 'Selective acculturation and the dating process: The patterning of Chinese-Caucasian interracial dating', *Journal of Marriage and the Family*, 32, pp. 273–8.

YU, L.C. (1984) 'Acculturation and stress within Chinese American families', *Journal of Comparative Family Studies*, 15, pp. 77–94.

Chapter 12

Mathematics and New Zealand Maori Girls

Sharleen D. Forbes

Introduction

New Zealand was one of a number of countries that participated in the 1981 SIMS (Second International Mathematics Study) mathematics survey. In certain areas of mathematics, such as measurement and arithmetic, New Zealand students failed to attain a high international ranking (Department of Education, 1987). Moreover, small but statistically significant gender differences in overall mathematics performance were observed as early as Form 3 (the first year of secondary school) (Wily, 1986). In comparing ethnic groups within New Zealand, marked differences which, in general, increase throughout the Form 3 year (Garden, 1984) were noted between Maori (indigenous New Zealanders) and other students. Data for the next international survey, the Third International Mathematics and Science Study, is not due to be collected in New Zealand until late 1994. A smaller national study was therefore commissioned by the Ministry of Education and undertaken by the women's collective, Equity in Mathematics Education, using 1989 data to update some of the SIMS results.

The Equity in Mathematics Education study involved a 9 per cent sample of all first-year secondary students in one of the four education regions that comprise New Zealand. While Maori students comprised only 17 per cent of the first-year secondary school population in 1989 (Ministry of Education, 1990), they were slightly over-represented (23 per cent) in the sample. In March, the beginning of the school year in New Zealand, a sample of 1639 students completed questionnaires focusing on their family background and attitudes towards mathematics. The sample was designed in such a way that approximately half of the students were from 'intact classes', that is a class consisting of all students taught mathematics together at the same time and by the same teacher. Tests from the SIMS survey were administered to students in March, and again at the end of the school year in November, to permit an assessment of the impact on students of the mathematics programme. The purpose of this chapter is to investigate the mathematics achievement of Maori girls, and to determine whether the observed overall gender differences in mathematics achievement in New Zealand are culture-specific. Since the average age of the students in the study was 13 years, this report provides an analysis of the attitudes, expectations and mathematical outcomes of 13-year-old Maori girls. For full details of the study including a comparison of the overall, topic, and question-specific results between 1989 and 1981, and an analysis of gender and ethnic differences in attitudes towards, and performance in, mathematics, see Forbes *et al.* (1990).

Backgrounds and Attitudes of Maori Students

Under the Treaty of Waitangi, the Maori people have special rights with respect to their land, fisheries, and forests. However, since signing the agreement in 1840, they have never experienced the same social and economic advantages enjoyed by other sectors of New Zealand society.

Maori is not the first language of most Maori children. While approximately 43 per cent of the 384 Maori students in the study said that their parents normally spoke the Maori language at home, only 10 per cent of the students claimed to be able to speak it themselves. Perhaps as a result of deliberate attempts by educators to 'assimilate' Maori students, English has become their dominant language and is used almost exclusively in everyday conversation. The preservation of the Maori language is currently a major concern in New Zealand, with many Maori parents beginning to demand bilingual or immersion Maori education for their children (Davies and Nicholl, 1993).

The occupations of employed parents were classified using the Elley-Irving (Johnson, 1985) scale for males and the Irving-Elley (1977) scale for females. The 'unskilled' category includes manual labourers, but not home-makers, the unemployed, or those on welfare. As Table 12.1 shows, parents of Maori students are much more likely than parents of European descent to be in 'unskilled' or 'semi-skilled' occupations (categories 1 and 2 in Table 12.1).

Maori parents were also more likely than their European counterparts to have no education beyond the secondary level (Table 12.2). Less than 20 per cent of both Maori mothers and Maori fathers participated in tertiary education, compared with approximately one-third of parents of European descent. There were reasonably high positive correlations between the employment and education variables ($p < 0.05$), indicating that a parent with limited education would most likely be in an unskilled occupation, with a spouse in a similar position. Not only do most Maori children come from low-income families, but as already noted, their parents are also likely to possess only a limited education. This might be expected to affect the attitudes of Maori children towards higher education, as well as to hinder their participation in certain subject areas such as mathematics.

The disadvantages experienced by Maori students in terms of their low socio-economic status, coupled with the discontinuity they experience between their home and school cultures, influence their attitudes towards post-secondary education. Only 32 per cent of the Form 3 Maori students expected to continue

Table 12.1: The percentage of parents of European and Maori origin in each socio-economic category

	Fathers		Mothers	
	European	Maori	European	Maori
1	7	23	16	27
2	12	34	10	18
3	34	24	25	22
4	23	14	33	24
5	12	2	11	7
6	12	3	5	2

Note: Scale: 1 = Unskilled, 6 = Professional

Table 12.2: Highest level of education of parents (by percentage)

	Fathers		Mothers	
	European	Maori	European	Maori
Primary only	2	5	2	5
Secondary only	63	80	67	77
Tertiary	35	15	31	18

with formal education beyond secondary school, compared with 57 per cent of students from European backgrounds. There was a significant correlation ($p < 0.05$) between the students' expected number of years of further education and the parents' education levels or occupations, indicating that these low expectations are, at least in part, influenced by home characteristics.

The March questionnaire asked students to indicate the level of their agreement or disagreement with the following statements on a five-point scale ranging from strongly disagree to strongly agree:

- Mathematics is useful in solving everyday problems.
- It is important to know mathematics in order to get a good job.
- Most people do not use mathematics in their jobs.
- Boys have more natural ability than girls.
- Boys need to know more mathematics than girls.
- If I had my choice I would not learn any more mathematics.

For the purposes of this chaper, the two 'disagree' and two 'agree' categories have been combined. A high percentage of students in both ethnic and gender groups (80 per cent) agreed that mathematics is useful for solving everyday problems, and that it is important to know mathematics in order to get a good job (85–90 per cent). The majority of Maori boys and girls (58 and 60 per cent, respectively) believed that most people use mathematics in their jobs, but these percentages were significantly less ($p < 0.05$) than those for boys and girls of

Table 12.3: Students' perceptions of gender differences in mathematics (by percentage)

	Boys have more ability in maths				Boys have more need for maths			
	European		Maori		European		Maori	
	M	F	M	F	M	F	M	F
Disagree	62	94	57	86	74	86	58	63
Undecided	29	5	30	13	19	9	26	27
Agree	9	1	13	1	7	5	16	10

European descent (71 and 77 per cent, respectively). A higher percentage of Maori girls were 'undecided' about whether most people use mathematics in their jobs than students in any other group.

For both groups of students, significantly more girls than boys disagreed that boys possess more natural mathematics ability or greater need to know mathematics than girls (see Table 12.3). Maori boys were the group most likely to agree with these statements, while significantly more Maori girls than girls of European descent were 'undecided' in this area. Further investigation will be necessary to determine the extent to which the strong views held by Maori boys influence those of Maori girls.

Mathematics is a compulsory subject for the first nine years of schooling, that is until the end of the second year in secondary school. Even in the tenth year, almost all students take mathematics (97 per cent in 1990), with a large majority (75 per cent in 1990) continuing to study the subject the following year (Ministry of Education, 1991). Given this, it was expected that a high percentage of students in all groups, on beginning secondary school, would state that they would continue with mathematics. Only for Maori boys was there a relatively large percentage who had already decided they would drop the subject (13 per cent compared with 5–6 per cent in other ethnic groups). However, a high proportion of Maori girls were 'undecided' (21 per cent compared with 10–13 per cent in other groups).

In summary then, at the beginning of secondary school, Maori girls as a group were the most likely to be 'undecided' about the usefulness of mathematics, whether boys had more natural mathematics ability or need for mathematics, and whether they themselves would continue to study mathematics when given the choice. Particularly disturbing is the observation that, upon entry to secondary school, the majority of all Maori students had already decided that they had no future in education beyond the secondary level.

Mathematics Performance

Performance was analysed for those students in the sample who sat the March and the November tests. The tests administered in March and November were

Table 12.4: Gain in test performance (March–November, 1989) by gender and ethnicity

	N	Mean	Standard Deviation	Median	Mode	Max	Min
European							
Males	302	3.79	4.81	4	4	7	−12
Females	265	4.54	4.47	4	4	17	−9
Maori							
Males	83	3.55	4.39	3	3	21	−5
Females	79	2.95	4.37	3	3	14	−10

Note: Numbers do not add to 813 since only non-Maori students of European descent are included.

identical, and were scored out of forty points. The mean score for the entire sample in March was 15.45 (with standard deviation 7.25), while the corresponding November figure was 19.51 (standard deviation 8.35). In other words, this is a mean improvement during the school year of only 4.06 marks, or 10 per cent. This small gain and the accompanying large standard deviation, coupled with the observation that one in every six students showed no improvement at all, are cause for grave concern.

The situation is worst for Maori students. Not only do they enter secondary school with a lower mean score than their European counterparts, they also exhibit significantly lower average gains in marks throughout the year. By November, the mean difference in score between Maori students and those of European origin was 5.7 marks in favour of the latter. Furthermore, although there is a positive correlation between parents' occupations or education levels and student performance, when these variables are controlled for, large differences in test achievement continue to remain between the two groups. Thus, the mathematics performance of Maori students cannot be explained simply by referring to their low socio-economic status, and it is quite possible that 'cultural oppression' has played a role here.

As illustrated by Table 12.4, Maori girls experienced the lowest gains in achievement during the school year. However, it is also the case that, when the mean scores of 'intact classes' were compared, one of the classes that witnessed the highest gain relative to its March score contained only Maori students of whom 60 per cent were female. This was a class of students in a small, rural, bilingual school with a young, male Maori teacher. Although the class had the lowest mean score in March, it gained an average of 4.6 marks over the course of the academic year. Given this, it seems plausible that a combination of teacher enthusiasm and appropriate learning environment could make a critical difference in outcomes for Maori students in learning mathematics.

Table 12.5 provides a comparison of the November 1989 test scores with those obtained by students in the SIMS survey (Department of Education, 1987). While in 1981 the gender difference in both groups was in favour of the boys, in 1989 we witness an interaction between gender and ethnicity, that is, only for Maori students is the mean score for girls lower than that of boys.

The improvement in the mean score for girls of European descent is

Table 12.5: November Test Performances by gender and ethnicity for the years 1989 and 1981

	Mean		**Standard Deviation**	**N**	
	1989	**(1981)**		**1989**	**(1981)**
European					
Male	20.7	(20.8)	8.59	304	(2076)
Female	21.1	(20.0)	8.19	267	(2037)
Maori					
Male	15.7	(15.9)	6.59	86	(310)
Female	14.6	(14.4)	6.19	80	(269)

Note: Numbers do not add to 813 since only non-Maori students of European descent are included.

encouraging for campaigns such as 'Girls Can Do Anything' (New Zealand's television and school promotion run by the Department of Labour in 1983), as well as for intervention programmes such as 'EQUALS' (1989) and 'Family Math' (Stenmark *et al.*, 1986), which have been actively promoting mathematics for girls. However, these endeavours do not appear to have had the same positive impact on the participation and performance of Maori girls. Perhaps this is not surprising, since the majority of the teachers and administrators involved in these programmes are not Maori themselves. With this in mind, it seems likely that Maori involvement and ownership of the development of approaches will be necessary before similar improvement will be witnessed in the mathematical participation and achievement of Maori girls.

Impact on Further Mathematics

Given that ethnic differences in mathematics performance are apparent at the beginning, and continue to the end, of the first year of secondary school, one might expect these differences to persist through the secondary years. Indeed, as we shall see, this is the case. In the normal course of events, the students in the 1989 sample would be in their third year of secondary school by 1991. It is in this year, Form 5, that the majority of students sit their first national examination, the School Certificate. Although mathematics is optional at this stage, it was the second most popular subject, after English, in the 1991 School Certificate examination for both groups of students under consideration. It is not, however, surprising, given the early differences in performance and attitudes of the students, that while mathematics was taken by 84 per cent of non-Maori School Certificate candidates who were in their third year of secondary schooling in 1991, it was taken by only 67 per cent of the corresponding Maori candidates.

Candidates who pass the School Certificate receive a grade ranging from D (lowest) to A1 (highest). In addition to the previously discussed differences in participation, there were also significant differences in achievement as shown in

Table 12.6: Percentage of 1991 School Certificate graduates in each grade category

	A1	A2	B1	B2	C1	C2	D	Number of Candidates
Maori								
Females	0.9	4.1	11.4	21.7	33.7	25.1	3.1	2688
Boys	1.5	5.3	14.1	24.1	30.9	20.9	3.2	2682
Total	1.2	4.7	12.8	22.8	32.3	23.0	3.2	5370
All	5.6	12.4	21.8	23.8	22.6	12.3	1.5	39394

Table 12.6. The most common grade for Maori students was C1, compared with B2 for non-Maori candidates (Ministry of Education, 1992).

This apparently bleak situation continued into the fourth and final year of secondary school, Form 6, with only 57 per cent of the Maori 1991 School Certificate mathematics candidates (59 per cent of girls and 52 per cent of boys) taking the Sixth Form Certificate in mathematics one year later. This stands in marked contrast to the 73 per cent figure for the corresponding group of non-Maori students (New Zealand Qualifications Authority, 1993).

When the small group of high-achieving (with grades of A1) Maori girls in 1991 School Certificate mathematics was examined more closely in terms of their performance in Sixth Form Certificate mathematics the following year, these girls scored at least as well as their female peers of European origin. The same, however, did not hold true for Maori boys (Forbes and Mako, 1993). This observation may indicate that the few Maori girls who continue with mathematics to this level are the very high-achieving. Despite this, throughout secondary school Maori girls, in general, exhibit the lowest participation and achievement levels in mathematics when compared either to Maori boys, or to New Zealand students of European descent.

Conclusion

To summarize, in this chapter, I have shown that on entry to secondary school, Maori girls are 'undecided' about the usefulness of mathematics, whether they should continue to study mathematics, and whether mathematics is a subject of greater importance to boys. Their average performance in mathematics at age 13 is lower than that of Maori boys, in contrast with the average performance of girls of European descent which equals that of their male peers. Furthermore, the observed low mathematics performance of 13-year-old Maori students compared with that of students of European descent increases throughout secondary schooling.

It appears that strategies to increase the participation and achievement of girls in mathematics in New Zealand are having a positive impact on girls of European descent, but are failing to meet the needs of Maori girls. The latter generally derive from low-income families, and commence secondary school with ambivalent views towards mathematics. However, these variables do not account for all the observed differences between Maori and other girls. Both the learning environment and the individual teacher may be critical factors in determining the success of Maori girls in mathematics. Maori girls belong to a culture that has suffered from misguided attempts at assimilation by the dominant group in New Zealand society, those of European descent. These attempts have culminated in a major loss of traditional Maori knowledge and language. It is likely that precisely targeted and Maori-driven strategies will be needed to increase the participation, achievement, and enjoyment in mathematics of Maori girls.

References

ELLEY, W.B. and IRVING, J.C. (1985) 'The Elley-Irving Socio-Economic Index, 1981 Census Revision', *New Zealand Journal of Educational Studies*, 20, 2, pp. 115–28.

DAVIES, L. and NICHOLL, K. (1993) *Te Maori i roto i nga Mahi Whakaakoranga: Maori in Education*, Wellington, Ministry of Education.

DEPARTMENT OF EDUCATION (1987) *Mathematics Achievement in New Zealand Secondary Schools: A Report on the Conduct in New Zealand of the Second International Mathematics Study within the International Association for the Evaluation of Educational Achievement*, Wellington, Department of Education.

EQUALS (1989) *Assessment Alternatives in Mathematics*, Lawrence Hall of Science, University of California at Berkeley.

FORBES, S.D., BLITHE, T., CLARKE, M. and ROBINSON, E. (1990) *Mathematics for All?*, Wellington, Ministry of Education.

FORBES, S.D. and MAKO, C. (1993) 'Assessment in Education: A Diagnostic Tool or a Barrier to Progress', Paper presented at the 1993 World Indigenous Peoples Conference: Education, Wollongong, Australia.

GARDEN, R.A. (1984) *Mathematics and Ethnicity*, Wellington, Department of Education.

IRVING, J.C. and ELLEY, W.B. (1977) 'A socio-economic index for the female labour force in New Zealand', *New Zealand Journal of Educational Studies*, 12, 1 and 2, pp. 154–63.

JOHNSON, R. (1985) *A Revision of Socio-Economic Indices for New Zealand*, New Zealand Council for Educational Research.

MINISTRY OF EDUCATION (1990) *Maori Education Statistics 1989*, Wellington, Ministry of Education.

MINISTRY OF EDUCATION (1991) *Education Statistics of New Zealand 1990*, Wellington, Ministry of Education.

MINISTRY OF EDUCATION (1992) *1991 New Zealand Senior School Awards and Examinations*, Wellington, Ministry of Education.

NEW ZEALAND QUALIFICATIONS AUTHORITY (1993) *Private Communication*, Wellington.

STENMARK, J.K., THOMSON, V. and COSSEY, R. (1986) *Family Math*, Lawrence Hall of Science, University of California at Berkeley.

TE PUNI KÖKIRI (1993) *Pängarau — Maori Mathematics and Education*, Wellington, Ministry of Maori Development.

WILY, H. (1986) 'Women in mathematics: Some gender differences from the IEA [International Association for the Evaluation of Educational Achievements] Survey in New Zealand', *New Zealand Mathematics Magazine*, 23, 2, pp. 29–49.

WONG, K. (1988) *Analysis of the Third Form Data from the 1981 IEA Mathematics Study*, Mathematics Honours Project, Wellington, Victoria University.

Chapter 13

Equal or Different?: Cultural Influences on Learning Mathematics[1]

Gilah Leder

Introduction

The contents of the influential *Handbook of Research on Mathematics Teaching and Learning* (Grouws, 1992) illustrate clearly the wide range of factors believed to influence mathematics learning. In addition to learner and classroom-related variables, reference is also made to the importance of the broad social context in which learning takes place. For example, the comprehensive model proposed by Eccles (1985) to explain students' decisions to enrol in mathematics courses (or opt out of them) includes the cultural milieu, the value-system adopted by, or attributed to, important socializers, as well as students' perceptions of the social climate in which educational and career choices are made.

The pervasive influence of external and societal factors is frequently examined when gender differences in mathematics learning are discussed (Leder, 1992). Listed among such factors are the influence of important others — teachers, parents, members of the peer group — as well as environmental and organizational influences. In this chapter, (print) media references to male and female roles, achievements, and behaviours are used as a measure of societal attitudes, beliefs and expectations, that is, of the general climate in which learning occurs and in which educational and career choices need to be made. The sources of interest are broader than the ones of earlier work (Leder, 1984, 1986, 1988) in which I concentrated on the ways successful females were portrayed in the print media in *one* country and discussed these accounts from a social-psychological perspective. This study focuses on print media portrayals in two ostensibly similar countries — Australia and Canada — and discusses these media images in two ways — in terms of feminist perspectives as well as the social-psychological approach (the expectancy-value theory of motivation) used in the earlier work. Implications of the cultural climate described for students' perceptions about mathematics and its relevance to them are also explored.

Social Context

Australia and Canada are geographically large and have comparatively small populations, approximately seventeen million, and twenty-seven million inhabitants, respectively. In both countries, the bulk of the population is concentrated in a small number of cities. Women comprise approximately 51 per cent of the population and just over 40 per cent (41.8 per cent in Australia, 40.8 per cent in Canada) of the labour force. Of female employees, 55 per cent in Australia, and 58 per cent of those in Canada are still concentrated in two major occupational groups: clerks and salespersons. Canadian statistics indicate that the proportion of women in these traditionally female occupations rose from 55 per cent in 1971 to 58 per cent in 1986. A comparable increase in such occupations over a similar period of time was reported for Australia by Sampson (1993). Data for 1991 show that approximately 7 per cent of Australian working women are in managerial and administrative positions. Canadian data for 1988 reveal that 10 per cent of the female workforce in that country are in such positions. Possible differences in job classification procedures make comparisons such as these problematic, however. In both countries, women are much more likely than men to hold part-time jobs: some 40 per cent and 25 per cent of all female employees in Australia and Canada respectively compared with one in ten working males in both countries.[2]

The Print Media

The media are generally regarded as powerful reflectors and determinants of behaviour (Eysenck and Nias, 1978; Faludi, 1991; Jacobs and Eccles, 1985). Jacobs and Eccles (1985) reported that parents' beliefs, in general and for their own children, about gender differences in mathematics learning became more stereotyped after they were exposed to media coverage of a heated debate about the more likely reasons (environmental or genetic) for gender differences in mathematics performance. 'Media absorption', they argued, 'involves self-selection according to previous attitude' (Ibid., 1985, p. 24).

While the impact of television is outside the scope of this chapter, it is nevertheless of interest to note that the long-running series, *Sesame Street*, often cited as an effective enrichment programme for socially disadvantaged children, 'has achieved a special status as part of the cultural milieu from which children are thought to acquire a wide range of early childhood competencies — early number competencies among them' (Secada, 1992, p. 641).

There is often concern about the manner in which issues are reported by the media. Faludi (1991) stressed the implications of biased reporting on women's lives:

> The Victorian era gave rise to mass media and mass marketing — two institutions that have since proved more effective devices for constraining women's aspirations than coercive laws and punishments. They rule with

> the club of conformity, not censure, and claim to speak for female public opinion, not powerful male interests. (Faludi, 1991, p. 48)

She further argued that incalculable harm was being done by contemporary media stories:

> The press was the first to set forth and solve for a *mainstream* audience the paradox in women's lives . . . women have achieved so much yet feel so dissatisfied; it must be feminism's achievements, not society's resistance to these partial achievements, that is causing women all this pain. (Ibid., p. 77, my emphasis)

That journalists themselves are aware of the limitations of their own reporting or that of their colleagues is made explicit from time to time. In an article published in a Canadian newspaper, the *Globe and Mail*, headed in part 'on the sins and omissions of the anglo press', one columnist described as biases of the English-Canadian media in their coverage of Québec and constitutional issues: blatant distortions, gross mistakes and flawed analysis or knee-jerk stereotyping (Gagnon, 1992, p. D3).

A perceived need to examine how issues pertaining to gender equity, to women's roles and lifestyles are portrayed in the popular press in two apparently similar countries, Australia and Canada, served as the stimulus for this contribution. Such a review should reveal the broad cultural climate in which students learn and in which they need to decide — among other issues — whether or not to strive for success in mathematics and whether to continue with mathematics once the subject is no longer compulsory. Such decisions have important implications for long-term career and life opportunities. The ways in which mathematics qualifications act as a critical filter to tertiary courses and apprenticeships have been well documented (Sells, 1973; Tobias, 1993).

The Study

Two newspapers with a large circulation and readership were monitored over a two-week period in April and May of 1992: the *Globe and Mail* in Canada and the *Age* in Australia. Articles about gender-equity issues, about women's roles, and ones that contrasted female and male lifestyles were of particular interest. The *Globe and Mail* was monitored independently by a second reader. Interrater agreement was 94 per cent. Only articles selected by both readers are included in this report.

Once selected, a content analysis was made of each of the articles using two distinct methods.

The Social-Psychological Approach

The expectancy-value theory of achievement motivation has been widely used to explain and predict behaviour (McClelland *et al.*, 1953). Within this framework, it is argued that individuals are more likely to strive for goals which they value and believe they can attain. A description of an occupation as male-dominated and 'uncomfortable' or difficult for females affects both the value and perceived likelihood of attaining that occupation. Educational paths selected, or advised, (for example, participation in mathematics and related careers) are likely to be shaped by these perceptions (Eccles, 1985).

Based on previous media searches, a number of common themes were anticipated: that females are disadvantaged — though the fields in which disadvantage might occur were not predicted; that male and female roles continue to be different; and fear of success imagery. Relevant to this last category were occurrences where success was described as being attained through luck or without effort, instances where success in a career was said to have been achieved at a personal cost, confirmations that it was more difficult for females than males to be successful, and gratuitous references to interpersonal relationships or significant others.

Feminist Perspectives

Three major themes were anticipated, conveniently designated as liberal feminisms, social feminisms, and feminisms of difference (compare with Mura, Chapter 19, this volume). Given recent explorations of links between feminist pedagogy and mathematics (for example, Damarin, in press; Jacobs, 1994), an examination of print-media publications for feminist themes seemed timely.[3] Jacobs (1994), for example, argued that 'aspects of feminist pedagogy can be used to make women more willing to study mathematics, and therefore learn mathematics' (Jacobs, 1994, p. 12).

Articles which included explanations of gender imbalance in terms of barriers that restricted women's access to current social benefits and particular sectors of the labour market were deemed to be consistent with liberal feminisms if their content implied that the imbalance could be eliminated by the removal of barriers and the re-socialization of women. In such articles it was often argued that females should be given equal opportunities to access the elements of the current social system which are defined within that system as positive and desirable.

Articles which pointed to a gender imbalance and explained this as a result of the contemporary social structures, rather than women's failure to fit into them, were considered consistent with social feminisms. Here the current social structure was regarded as a mechanism of domination, oppression and exploitation of females so that the structure itself would need to be changed before women as women would be able to participate successfully and willingly in all spheres.

Items with a theme that gender imbalance was the result, in our culture, of the centrality of masculine values and could be redressed only if the positive dimensions of women's nature were identified and promoted, were considered as representative of feminisms of difference. In this category experiences, desires and spirituality of females would be regarded as essentially different from those of males.

The three broad feminist perspectives outlined above are not mutually exclusive. There is, in particular, an overlap between social feminisms and feminisms of difference because both understand women to have different needs and experiences from those of men. An important difference between these two perspectives is the social feminists' view that the reform of society required to include female needs and desires would be for the good of both women and men, and that feminine qualities are not necessarily women's qualities, that is, gender differences have a social origin. Feminisms of difference, on the other hand, are separatist in that they understand these qualities to be essentially female and, therefore, that females need different social structures, not only from the current ones, but from the ones of males. Both social feminisms and feminisms of difference criticize the liberal approach because it identifies women and femininity as the problem and implies that women need to be given the opportunity to be more like men if they are to succeed.

The argument that females' experiences and subjectivity are different from those of males is common to all feminisms. It is the perceived origin of this difference that distinguishes between them: exclusion of women from legitimized realms of society for liberal feminisms; the oppressive and exploitative nature of legitimized society and of the laws that legitimize that structure for social feminisms; and the need to recognize, accept, value, and legitimize essential differences for feminisms of difference. All three perspectives point to directions in which mathematics might be reconceptualized and new strategies be introduced into the mathematics classroom (Damarin, in press; Jacobs, 1994; Mura, Chapter 19, this volume).

Results

Articles dealing with gender-equity issues or highlighting successful women appeared in each newspaper for each day the papers were monitored. Their length and frequency varied. Some typical examples are shown below.

- **Example 1**
  - Headline: 'A prima donna from the start'
  - Summary: A Swedish soprano had an incredible debut and became a star overnight. 'It was pure luck', she recalls. Her Austrian husband is a baritone. Her sons are now old enough for her not to worry when she goes on tour. (Setting: success in music)
  - Coding: 1. Fear of success imagery: success was 'pure luck', gratuitous reference to others.
    2. No evidence of feminisms theme.

- **Example 2**
  - Headline: 'Getting mixed messages on being a woman'
  - Summary: A series of articles which convey conflicting messages about the

role of contemporary women is summarized. 'Every working woman with children knows what it means to be caught between conflicting hopes for herself and the expectations of family.' 'Even if such things as good day care, understanding employers and supportive husbands were universally guaranteed to working mothers, I'm not sure that women could ever fully avoid the "discontinuity" of a mother's life.' 'Men are still calling for the shots.'

Coding: 1. Fear of success imagery: conflict between career and motherhood needs.

2. i) Social feminisms: evidence of imbalance in contemporary social structures;

ii) Feminisms of difference: reference to difficulty of 'equalizing the roles of men and women', given the centrality of masculine values.

- **Example 3**

Headline: 'Plans for tougher affirmative action'

Summary: Tougher penalties should be imposed against employers who discriminate against women in the workplace. Smaller companies should no longer be exempt from the Affirmative Action Act. Many companies have 'done very little to improve opportunities for women'.

Coding: Liberal feminisms: gender equity (deemed to be desirable) has not been achieved in the workplace but better legislation should lead to improvements.

- **Example 4**

Headline: 'Former dancer keeps House in order'

Summary: Miss Boothroyd 'made history' when she became the first woman to be elected Speaker of Britain's House of Commons. As one of sixty women in the 651 member House she had earned a reputation for fairness and impartiality as the Deputy Speaker. She is 'a former chorus-line dancer and secretary'. (Setting: success in politics)

Coding: 1. Fear of success imagery: 'It will be a lonelier life than I have ever known before.'

2. Liberal feminisms: Previous experience (her record as Deputy Speaker) indicates that she is able to overcome real and/ or perceived barriers and to compete successfully in a traditionally male domain. Her performance has been, and is likely to continue to be, judged by male standards.

- **Example 5**

Headline: 'Women coming to grips with math ghost'

Summary: Like other students, female student teachers often find mathematics more difficult than their male peers. Many universities have now introduced courses 'explicitly aimed at eradicating math phobia'. Schools are also more aware that girls need to be encouraged in mathematics classes.

Coding: 1. Perpetuation of stereotype: Females often have more difficulty than males with mathematics.
2. Liberal feminisms: Females should aim to perform equally with males. With appropriate assistance, there need be no differences in the performance of females and males (in mathematics).

Discussion

Both newspapers contained articles with fear of success imagery and feminist themes. A number of these focused on the fate of females generally. In others the metaphors were introduced in stories about specific individuals. The lack of progress made towards gender equity was discussed in several articles.

Both newspapers contained articles which confirmed that females were widely disadvantaged — in politics, in the arts, in the workforce generally, in jail, in the lifestyle thrust upon them. Several of these stressed differences between groups while ignoring differences within groups. Interestingly, males were described as being disadvantaged in one (Canadian) article, females as being specifically advantaged in two (Australian) articles. One of the latter described females' higher pass rates at senior high-school level in a wide range of subjects.

In both newspapers there were descriptions of affirmative action practices, with 'Women fighting back' the dominant theme in two of the (Canadian) articles.

Comparisons between gender and race discrimination were made in several of the (Canadian) articles. During the two-week period monitored, this more complex issue did not attract the attention of the Australian media.

The daily appearance of articles which confirmed traditional stereotypes, dwelt on the difficulties faced by females, stressed family–career conflicts experienced by them, focused on societal pressures and/or the prevalent acceptance of male standards and values is noteworthy. While the frequency with which they appeared, as well as their placement in the paper (rarely on the front page) should be balanced against the large number of other issues discussed each day, their impact should not be underestimated. A further, intensive testing of Jacobs and Eccles' (1985) conclusion that 'media absorption involves self-selection according to previous attitude' (p. 24) is clearly warranted. Identifying, discussing, and exploring the origins of the areas of disadvantage depicted (for example, unequal opportunities for participation in selected areas, social or organizational barriers, the [in]appropriateness of using male behaviours or standards as the norm) may serve to counteract negative images and point to constructive initiatives for the mathematics classroom.

Concluding Comments

Gender differences in mathematics learning have attracted much research attention. Reference is frequently made in this work to the influence of the social climate and environmental factors. An examination of the contents of two large and widely read metropolitan daily newspapers published in two different, highly industrialized countries, revealed the regular inclusion of articles which more frequently confirmed, than challenged, established male values and power structures, and dwelt on traditional stereotypes and the difficulties faced by females in particular in juggling professional and personal pressures. The impact of these messages on adolescents and their parents requires careful monitoring. Meanwhile, it appears that the subtle messages conveyed in the popular press are consistent with small but consistent differences in the ways females and males perceive and value mathematics and related careers as appropriate for themselves.

Notes

1 An earlier version of this chapter was presented at the IOWME Study Group held as part of the ICME-7 in Québec City, Canada, August 1992.
2 Sources: *Women at Work* (1991) 13, 3, Australia, pp. 12–13 and *1992 Canada Year Book*, Ottawa, Statistics Canada.
3 The contribution of Julianne Lynch to this analysis is gratefully acknowledged.

References

DAMARIN, S.K. (in press) 'Gender and mathematics from a feminist standpoint', in SECADA, W.G., FENNEMA, E. and BYRD, L. (Eds) *New Directions in Equity for Mathematics Education*, Cambridge, Cambridge University Press.

ECCLES, J.S. (1985) 'Model of students' mathematics enrolment decisions', *Educational Studies in Mathematics*, 16, 3, pp. 311–14.

EYSENCK, H.J. and NIAS, D.K. (1978) *Sex, Violence and the Media*, New York, Harper and Row.

FALUDI, S. (1991) *Backlash*, New York, Crown.

GAGNON, L. (1992) 'On the sins and omissions of the anglo press', in DAFOE, J. and LA PIERRE, L. *Globe and Mail*, May 9, Section D3.

GROUWS, D.A. (Ed) (1992) *Handbook of Research on Mathematics Teaching and Learning*, New York, MacMillan.

JACOBS, J.E. (1994) 'Feminist pedagogy and mathematics', *Zentralblatt für Didaktik der Mathematik*, 26, 1, pp. 12–17.

JACOBS, J.E. and ECCLES, J.S. (1985) 'Gender differences in math ability: The impact of media reports on parents', *Educational Researcher*, 14, 3, pp. 20–4.

LEDER, G.C. (1984) 'What price success? The view from the media', *The Exceptional Child*, 31, pp. 223–4.

LEDER, G.C. (1986) 'Successful females: Print media profiles and their implications', *The Journal of Psychology*, 120, pp. 239–48.

LEDER, G.C. (1988) 'Fear of success imagery in the print media', *The Journal of Psychology*, 122, pp. 305–6.

LEDER, G.C. (1992) 'Mathematics and gender: Changing perspectives', in GROUWS, D.A. (Ed) *Handbook of Research on Mathematics Teaching and Learning*, New York, MacMillan, pp. 597–622.

MCCLELLAND, D., ATKINSON, J.W., CLARKE, R.A. and LOWELL, E.L. (1953) *The Achievement Motive*, New York, Appleton, Century, Crofts.

NEW SOUTH WALES ANTI-DISCRIMINATION BOARD (1984) *Discrimination and Religious Conviction*, Sydney, NSWADB.

SAMPSON, S.N. (1993) *Dismantling the Divide*, Canberra, Australia, Department of Employment, Education and Training.

SECADA, W.G. (1992) 'Race, ethnicity, social class, language and achievement in mathematics', in GROUWS, D.A. (Ed) *Handbook of Research on Mathematics Teaching and Learning*, New York, MacMillan, pp. 623–60.

SELLS, L. (1973) *High School Mathematics as the Critical Filter in the Job Market* (ERIC No. ED080 351), Berkeley, University of California.

TOBIAS, S. (1993) *Overcoming Math Anxiety*, revised edition, New York, W.W. Norton and Co.

Chapter 14

Gender Inequity in Education: A Non-Western Perspective

Saleha Naghmi Habibullah

In this chapter, I challenge the popular notion that western strategies for overcoming gender inequity in education are likely to be equally effective in non-western societies. In my opinion, there is a direct link between a country's socio-cultural conditions and the manner in which its people perceive and react to the issue of gender imbalance. For this reason, there exist enormous differences between the western world and many of the developing countries regarding this issue. Consequently, for many years to come, short-term goals and strategies for educational improvement will have to vary across countries in accordance with economic conditions and socio-cultural realities. In this regard, I suggest that a broad-based programme of research may prove to be very beneficial.

The issue of gender inequity in mathematics education must be considered in conjunction with the broader issue of gender imbalance in education as a whole. In this regard, it is important to understand the differences between the educational contexts in the underdeveloped and the developed worlds. From a purely theoretical viewpoint, one of the primary goals of education is to foster the development of the human mind so that individuals are able to make appropriate choices, succeed in professional and familial spheres, and indulge in a variety of creative activities in accordance with their inherent potentialities. While there is little doubt of this being an ideal that every country should aspire to in the long run, the sheer heterogeneity of today's world requires flexibility, thoughtfulness and careful consideration of what *is* and what is *not* feasible in a particular culture. Although we may overlook differences in the history, culture, and socio-economic conditions within each of the developed and underdeveloped realms, the disparities that exist between them cannot be ignored. Numerous facilities that are taken for granted in affluent societies may be scarce or absent in poorer nations. Furthermore, social norms considered appropriate in many non-western countries may be deemed completely unacceptable in western culture and vice versa. Taking these observations into account, it is clear to me that short-term educational goals must vary across countries in accordance with local cultural factors and socio-economic conditions.

The particular characteristics of each country, such as the socio-economic and cultural factors alluded to above, have a direct bearing on the manner in which its people perceive and react to the issue of gender inequity. As such, gender inequity in education does *not* present similar concerns in all parts of the world. A number of arguments can be employed to elucidate this observation. First, in countries where the procurement of basic necessities is the primary concern, gender-equitable education has a much lower priority. In such countries, for possibly many years to come, it may be only fair to focus the limited resources on poverty reduction, rather than on gender equity in education. Similarly, and this brings me to my second point, in countries with high illiteracy rates, and in some cases considerably higher illiteracy rates among women than among men, it is probably only fair to make reducing illiteracy a matter of top priority. Only after this has been achieved, can resources be devoted to more subtle issues such as reform of the male-oriented primary and secondary-school mathematics curriculum.

Third, in countries where, for religious or historical reasons, the male adult is accepted as the bread-winner for the family, pursuit of employment is a much greater concern for men than for women. Given the link between advanced education and higher paid jobs, it is not surprising that, in such countries, boys are more likely to aspire towards higher education. In such societies, efforts to reduce gender bias at the primary and secondary levels of education may prove to be more successful than attempts to reduce gender imbalance in higher education.

Finally, in societies where, for religious or cultural reasons, boys and girls receive their education at separate schools and colleges, the manifestation of gender inequality might be quite different from that in coeducational situations. At the secondary-school level, girls in sex-segregated classes probably enjoy a greater degree of self-confidence and freedom of expression than those in coeducational environments who might feel intimidated by the presence of boys who are generally more assertive. At the post-secondary level, however, lack of communication with male counterparts may reduce girls' exposure to situations outside of the domestic world and, as a result, hinder their intellectual development.

The unifying thread of the above discussion is the fact that historical, religious, socio-economic, and cultural realities all combine to shape a country's educational system as well as the mode of behaviour, the value-system, and the aspirations of its people. As previously mentioned, the phenomenon of gender inequality in education does not manifest itself in an identical fashion around the globe, and there is enormous diversity in how different peoples perceive and react to the issue. I have attempted to show that it would be impossible to develop a uniform strategy for educational improvement in all parts of the world. While the equitable development of humanity should remain the long-term universal goal of education, short-term educational objectives for a particular region of the world will necessarily have to be in accordance with local needs.

To this point, the western movement for gender equity in education has not given adequate recognition to the aforementioned realities. In order to develop a broader, less westernized view regarding the issue of gender imbalance, to widen people's understanding of cross-cultural differences, and to formulate appropriate strategies for educational improvement in different parts of the world, it is of the utmost importance that we explore and study in detail the socio-economic and cultural conditions of various countries. Such studies will necessarily have to focus, not only on economic conditions and educational and employment structures, but also on socio-cultural norms including people's beliefs, value-systems and aspirations. Comparative studies may prove to be extremely useful in ascertaining differences and similarities between nations, in determining the benefits and drawbacks of a variety of educational systems, and in the formulation of strategies for educational improvement in various parts of the world. In order to accomplish this objective, scholars from around the world should collaborate in the creation of a comprehensive, broad-based programme of research.

Chapter 15

Gender and Mathematics: The Singapore Perspective

Berinderjeet Kaur

Introduction

Over the last two decades, a major focus of educational research has been the investigation of sex-related differences in learning mathematics. Gender differences in mathematics education have been shown to be complex and influenced by the interaction of a variety of social factors, and hence to vary considerably across cultures (Ethington, 1990; Hanna, Kündiger and Larouche, 1990). In the case of Singapore, only a handful of studies to date (Chung, 1985; Kaur, 1987; Leuar, 1985; Ministry of Education, 1988; Tan, 1990; Tanzer and Sim, 1991) have contributed to the growing body of literature in this area. This chapter presents the Singapore perspective on gender issues in mathematics education through a review of the available Singapore research and discussion of their findings.

Review of the Literature

Two separate studies by Chung and Leuar, both in 1985, examined the relationship between gender and mathematical ability. Each study used a sample of first-year junior college (age 17 years) students selected on the basis of their mathematics grades in the GCE O-level (General Certificate in Education Ordinary Level) examination, a national examination written by all students at the end of their tenth year of formal schooling (age 16 years).

Chung (1985)

In her study, Chung examined the differences in attitudes and expectations of eighty students, of whom fifty had O-level mathematics scores of 70 per cent or more (the 'high' group), and thirty had scores between 50 and 59 per cent (the 'low' group), with equal numbers of males and females in both groups. The assessment instrument employed was a questionnaire comprising thirty-eight items adapted from the International Project for the Evaluation of Educational Achievement (Husén, 1967) and the Sandman (1979) Mathematics Attitude Inventory.

All respondents in the study viewed mathematics as an important subject, useful in the workplace as well as in society at large. While individuals in the 'high' group considered mathematics to be more important than those in the 'low' group, there were no significant gender differences. Males were found to be more confident than females of their ability to learn mathematics, with subjects in the 'high' group demonstrating a greater degree of confidence than their 'low'-ability counterparts. In terms of enjoyment, again there were no significant gender differences, but students in the 'high' group enjoyed mathematics more than those in the 'low' group. Only male members of the 'high' group believed in the common stereotype of mathematics being a male domain and, other than the fact that males in both groups preferred trial and error methods of problem-solving, no significant gender differences were found in students' views of the nature of mathematics. Some differences were found in students' career aspirations however. Although more females than males indicated an interest in accounting, a majority of male students indicated an interest in studying engineering at university, and teaching appeared to be the predominant career preference for females.

Leuar (1985)

Leuar's investigation was confined to 163 students (ninety-one boys, seventy-two girls), all of whom had obtained good results in O-level mathematics and were currently studying two advanced mathematics courses. Again, a questionnaire was employed. The issues investigated included: the extent to which students like mathematics compared with other advanced subjects studied; students' opinions as to whether girls or boys perform better in mathematics, science, and English language comprehension and communication; students' views on the usefulness and practicality of mathematics; and, whether students required the assistance of a private mathematics tutor.

Data from the study indicated that, although their levels of achievement in O-level mathematics were similar, males significantly outperformed females on the mathematics test taken at the end of the first year of junior college. Both sexes displayed similar patterns of affinity towards mathematics and both believed that males would do better in mathematics, although the boys surmised that girls would fare better in English language comprehension and communication. Regardless of sex, a majority of the subjects believed in the practical utility of mathematics, but did not know whether mathematics would be useful to them in their future studies. Very few of the study's participants had a private tutor, and most worked independently with occasional help from friends and teachers.

Kaur (1987)

In her study of gender differences in mathematics attainment, Kaur analysed performance on specific topics in the 1986 Singapore-Cambridge O-level examination in mathematics. The examination consisted of two papers, each of which assessed students' understanding of basic mathematical concepts and applications, as well as their ability to demonstrate this clearly in writing. The examination questions were classified by topic, level of cognitive complexity, and the degree of spatial ability required, and the points earned for each part of a question on both papers were recorded for use in data analysis. From a population of 42 627 examination candidates, with girls slightly outnumbering boys, a random sample of 176 was drawn with equal representation of the sexes from each of six achievement levels.

Overall, boys significantly outperformed girls on the compulsory questions. More specifically, boys did significantly better than girls on the following compulsory topics: mensuration, statistics, arithmetic, geometry, and probability. Girls, on the other hand, excelled in the areas of algebra and graphing. While boys surpassed girls on the questions testing spatial ability and cognitive complexity, no significant gender differences were found in the areas of computation and problem-solving. On optional questions, girls preferred problems requiring algebra, graphs, and 2-dimensional vectors, while boys' only marked preference was mensuration.

Ministry of Education (1988)

The Ministry of Education's study, as reported in Yip and Sim (1990), focused on the relationship between examination performance, gender, and birth-month for primary students who sat the PSLE (Primary School Leaving Examination) during the period 1977 to 1987. The findings indicated that at the age of 9 years, girls outperformed boys in their first language, usually English, their second language, usually the mother-tongue, and in mathematics. Students born earlier in the year also performed better than those born later the same year. By the time they sat the PSLE at 12 years of age, girls had maintained their superiority over the boys in the two languages, while the boys outperformed the girls in mathematics and science. Once again, those born earlier in the year seem to enjoy an advantage in terms of enhanced performance.

Tan (1990)

The focus of Tan's study was on mathematics achievement, mathematics anxiety, and locus of control. More specifically, he focused on these areas in the context of differences between males and females, and between arts and science students. The participants in the study comprised 558 16-year-old students chosen from six secondary schools, with approximately equal numbers of students in each of the four gender-stream groups. The instruments used in the study were the Fennema-

Sherman Mathematics Anxiety Scale, and a Mathematics Achievement Locus of Control Scale and Mathematics Achievement Test, both developed by the researcher himself specifically for the study.

A stepwise multiple regression analysis was performed with gender, stream, mathematics anxiety, and locus of control as independent variables, and mathematics achievement as the dependent variable. The results of this analysis showed stream to be the best regressor, followed by locus of control and, thereafter, mathematics anxiety. Gender was not a significant regressor, implying that mathematics achievement was not linked to the sex of the subject. Furthermore, a two-way analysis of variance showed no significant interactions between gender and stream for each of the dependent variables, mathematics achievement, mathematics anxiety, and locus of control. There were, however, significant differences in all three dependent variables between arts and science students, with science students outperforming arts students in mathematics, exhibiting less mathematics anxiety and being more likely than arts students to attribute control to internal factors. Indeed, the correlation between locus of control and mathematics achievement was observed to be significantly stronger in the case of arts students.

Tanzer and Sim (1991)

Tanzer and Sim's study examined self-concept and achievement attribution of primary-school students. Eight schools were chosen at random to participate in the study, with the proviso that the sample be representative of the different types of institutions existing in Singapore. From these schools, a sample of 1145 students was drawn, with 41 per cent aged 10 years and the remaining 59 per cent aged 12 years, and roughly equal numbers of boys and girls. The students represented a diversity of socio-economic backgrounds and their ethnic composition was a replication of the actual population, that is, 82 per cent Chinese, 12 per cent Malay, 5 per cent Indian and 1 per cent from other ethnic groups. A battery of scales measuring self-concept, achievement attribution and test anxiety were administered to the students in their respective classes by their teacher.

Gender differences in self-concept and attributional style were found to be domain-specific and consistent with prevailing stereotypes. In the younger age group (10-year-olds), boys' self-concept in regards to mathematics and school in general was higher than that of the girls, but this difference decreased with age. In addition, the younger boys were more likely to attribute their success in mathematics to effort than the girls were. Older students (12-year-olds), more than younger students, and irrespective of their gender, attributed their success in reading and in mathematics to external factors, such as luck and ease of task. Girls more readily than boys believed that failure in mathematics was the result of low ability, and failure in reading and in mathematics were attributed to poor effort more by the older students than by the younger ones. In addition, the older pupils were likely to view such failures as more the result of external than internal circumstances.

Discussion

Examinations play a vital role in Singapore's educational system (Yip and Sim, 1990, p. 147). The studies conducted by both Kaur (1987) and the Ministry of Education (1988; Yip and Sim, 1990) utilized available national examinations, and therefore, were able to involve the entire cohort of students at the desired levels. In terms of overall achievement in mathematics, the findings of both studies concur with the results in the wider body of related literature. However, the finding of no significant gender differences in the area of computation and problem-solving (Kaur, 1987) is in conflict with research findings from other cultures (see, for example, Sabers, Cushing and Sabers, 1987; Martin and Hoover, 1987). A potential explanation for this may lie in the fact that the preparation of students for the GCE O-level examination involves a substantial amount of time being spent practising previous examination questions and that the problem-solving questions included in the examination are quite routine.

What cannot, however, be achieved through a process of drill and practice is conceptual understanding of mathematics. At the advanced level of study in mathematics, where an in-depth understanding of the subject is vital, boys were found to outperform girls (Leuar, 1985). Improving girls' conceptual understanding of mathematics may, indeed, be the key to improving their mathematics achievement at the secondary level.

In contrast with Kaur's (1987) findings, Tan (1990), as previously mentioned, found no significant gender differences in mathematics achievement, mathematics anxiety, and locus of control among 16-year-olds. He suggests that the tenuous relationship between affective variables and achievement may be the result of the rapid modernization of Singapore society, and consequent emancipation of women. However, since his sample was not designed to be representative of 16-year-old Singapore students, generalizations should be treated with caution.

At present, there is no comprehensive study of Singapore students which might indicate the direct and indirect influences on achievement levels and orientations towards the learning of mathematics. In describing models of achievement behaviour, Ethington (1992) ascribes different paths to males and females. Support for this model is found in three of the studies reviewed here (Chung, 1985; Leuar, 1985; and Tanzer and Sim, 1991), which suggest that there may be different paths to understanding for male and female Singapore students.

Conclusion

At 16 years of age, boys outperform girls at O-level mathematics, a compulsory subject in Singapore schools. Given that a pass in mathematics is essential for entry to most tertiary institutions, why then do females appear to be lagging behind? Singapore is a traditionally male-dominated society in which mathematics and related skills are regarded as strictly male capabilities. Malathy (1992) argues that, as a result, girls tend to shy away from this 'masculine pursuit' and she suggests that the key to change may lie in the reform of teaching. In a fairly recent comprehensive review on mathematics education in Singapore, Khoo *et al.* (1991, p. 29) state that 'many questions remain to be explored in the area of gender and mathematics attainment, least of all, the true extent of any difference and the factors that may possibly account for the difference'. It is clear that more comprehensive, Singapore-based research will be necessary in order to assess the extent of any gender-related differences in mathematics education and develop strategies for redressing any observed imbalance.

References

CHUNG, Y.Y. (1985) 'Investigation of Sex and Ability-Based Differences in Attitudes and Expectations in Mathematics', Unpublished assignment during the period of induction, Singapore Institute of Education.

ETHINGTON, C.A. (1990) 'Gender differences in mathematics: An international perspective', *Journal for Research in Mathematics Education*, 21, 1, pp. 74–80.

ETHINGTON, C.A. (1992) 'Gender differences in a psychological model of mathematics achievement', *Journal for Research in Mathematics Education*, 23, 2, pp. 166–81.

HANNA, G., KÜNDIGER, E. and LAROUCHE, C. (1990) 'Mathematical achievement of grade 12 girls in fifteen countries', in BURTON, L. (Ed) *Gender and Mathematics — An International Perspective*, London, Cassell, pp. 87–97.

HUSÉN, T. (Ed) (1967) *International Study of Achievement in Mathematics — A Comparison of Twelve Countries*, New York, Wiley.

KAUR, B. (1987) 'Sex Differences in Mathematics Attainment of Singapore Pupils', Unpublished masters dissertation, United Kingdom, University of Nottingham.

KHOO, P.S., CHONG, T.H., FOONG, P.Y., KAUR, B. and LIM-TEO, S.K. (1991) *A State-of-the-Art Review on Mathematics Education in Singapore*, Singapore, Institute of Education.

LEUAR, B.C. (1985) 'Sex Differences and Mathematical Ability', Unpublished assignment during the period of induction, Singapore, Institute of Education.

MALATHY, K. (1992) 'Math — Must girls do worse than boys?', *Young Parents*, March–June, Singapore, Times Periodicals, pp. 32–7.

MARTIN, D.J. and HOOVER, H.D. (1987) 'Sex differences in educational achievement: A longitudinal study', *Journal of Early Adolescence*, 7, 1, pp. 65–83.

MINISTRY OF EDUCATION (1988) *Analysis of Performance by Sex on the PSLE 1977–1987*, Singapore, Ministry of Education.

SABERS, D., CUSHING, K. and SABERS, D. (1987) 'Sex differences in reading and mathematics achievement for middle school students', *Journal of Early Adolescence*, 7, 1, pp. 117–28.

SANDMAN, R.S. (1979) *Mathematics Attitude Inventory*, Minnesota Research and Evaluation Center, University of Minnesota.

TAN, O.S. (1990) 'Mathematics Anxiety, Locus of Control and Mathematics Achievement of Secondary School Students', Unpublished masters dissertation, Singapore, National University of Singapore.

TANZER, N.K. and SIM, Q.E. (1991) *Self-Concept and Achievement Attributions: A Study of Singaporean Primary School Students* (Research Report no: 1991/5), Institut für Psychologie der Karl-Frazens, Austria, Universität Graz.

YIP, S.K.J. and SIM, W.K. (1990) *Evolution of Educational Excellence: 25 Years of Education in the Republic of Singapore*, Singapore, Longman Publishers.

Chapter 16

Gender and Mathematics Education in Papua New Guinea

Neela Sukthankar

Introduction

At 14 years of age, Jenny is the oldest of six children. She serves as a secondary mother figure in her home, performing household duties such as cooking, cleaning, and looking after her younger siblings. She barely has sufficient time to go to school, let alone study enough to make attendance worthwhile. While this scenario may sound extreme, it is in no way atypical in a country like Papua New Guinea where the average family has five or six children. As in any other country, culture, social beliefs, and traditions all play an important role in the education of females. But, situations like the one described above serve only to widen the rift between male and female students. The aim of this chapter is to investigate the cultural issues affecting the education of females in Papua New Guinea and to suggest ways of improving the quality of their mathematical education.

Female Enrolment in the Education System

Papua New Guinea is a country comprised of countless, small, isolated communities with 700 dialects and numerous counting systems (Lean, 1987). Most people speak either Pidgin or Motu in addition to their village language. According to National Statistics Bureau data for 1988, 86 per cent of the population lives in rural areas where good schooling is hard to find. Children are forced to walk long distances in order to attend school and parents are reluctant to send their daughters under these conditions out of concern for their safety. While international schools with good educational facilities exist in the cities, they are far too expensive for the average salaried Papua New Guinean citizen to afford.

There is a substantial difference in the number of male and female students enrolled at the primary and secondary-school levels, with males outnumbering females. The gap increases at the tertiary level (Guthrie and Smith, 1980). This situation is in marked contrast to that in many developed countries where primary and secondary education is paid for through taxation and hence is mandatory. While having to pay to educate one's children does not create problems for the affluent, families with limited means are forced to make choices. In these circumstances, research has shown that families choose to educate their sons before their daughters. This is the situation in Papua New Guinea. In short, the issue can be linked to the level of education which the parents themselves have received. In families in which both parents were educated, it has been found that daughters enjoy better opportunities for schooling. Female students at a variety of scholastic levels in Papua New Guinea, including those participating in distance-education programmes, were found to have originated from backgrounds in which the parents were educated (Oliver, 1985).

Social and Cultural Influences

The social differences between men and women are reflected in the choices they make at school and university. Certain subjects such as woodwork, metal work, and the sciences are favoured by men, while others such as home economics, textiles, and child care are designated for women. Moreover, female students are expected to be docile and retiring in an atmosphere which generally equates boldness and flair with success. The ramifications of this are particularly serious in mathematics and science. Often the contradictions of being a women in science or mathematics are so overwhelming that females opt out in favour of arts-related subjects. Furthermore, since the learning potential of women is not valued, it is not encouraged. Unfortunately, this attitude is not confined to families, but is also institutionalized in girls' organizations such as the girl guides and brownies. In essence, these groups undermine the role of women in the workplace, by awarding badges for demonstrating skill mainly in such stereotypical female-roles as 'homemaker', 'childminder', and 'home nurse'. In this way, Papua New Guinean society places limitations on the spheres of activity in which a woman is permitted to develop a competitive spirit. These are a few of the ways in which social norms convey a different set of expectations for males and females, creating different lifestyles for each sex, with women viewed as subordinate to men. The good news is that there are signals that the situation is slowly improving in the cities where, as a result, the sex-related differences are gradually diminishing as well.

In Papua New Guinea society, a multitude of social and cultural customs combine to influence gender differences in education. At a broader level, it is also possible to identify certain customs that communicate information about sex-roles and sex-appropriate behaviour. One of these is the established system of 'bride price'. According to this practice, one which serves only to objectify women, a wife is 'purchased' from her parents by the bridegroom's family. Since a women's level of education is not valued it is not reflected in her 'price'. This custom is so pervasive that, even among well-educated families, the male child is provided with a good quality international education, while the female child is relegated to an inferior community school which continues to propagate the popular belief that a woman's life should be centred on her family, her home, and church affairs.

It should now be evident that any attempt to improve the quality of women's education must first and foremost focus on changing the society's attitudes towards women. These social attitudes which perceive women as inferior beings have, as already noted, been institutionalized by the society within its educational system. Admittedly, mathematics and science are considered to be 'male subjects' in many developed countries as well. But in Papua New Guinea, not only do males outperform females in these subject areas, they are expected to do so. Numerous remedial measures have been taken to combat the damaging attitudes that lead to such expectations. Although not common in Papua New Guinea, one measure that some countries are experimenting with is sex-segregated schooling. In Papua New Guinea, while most schools are coeducational, many provincial schools have single-sex classes in certain subject areas. For instance, it is normal practice for girls to study home science while boys do woodwork. While it appears that girls consistently perform better in these gender-specific classes, it is difficult to say how much of this can be attributed to segregation in itself since teachers expect girls to do well in these so-called 'female subjects'.

Mathematics Education

The majority of the population of Papua New Guinea, as noted above, lives in rural areas far removed from urban requirements of accounting, measuring and calculating. For the most part, then, the role of mathematics in people's daily lives is limited to elementary arithmetic. It is small wonder that children from rural backgrounds find it hard to accept the value of abstract mathematical ideas. Added to this, while it is true that women are officially encouraged to play an active role in the development of Papua New Guinea, the majority of them do not even have the opportunity to obtain a secondary education. Women pursuing higher-level mathematics in present-day Papua New Guinea are pioneers. Naturally, this means they face several obstacles, among them the fact that they have few female role models, their teachers being predominantly male. It is imperative, therefore, that any strategy designed to produce gender equity in mathematics education, should focus not only on encouraging more female students to take mathematics, but also on recruiting more female educators.

An associated problem arises because mathematics is the basis of fields such as engineering, the applied sciences, architecture and accountancy, all areas in which few Papua New Guinean women currently participate. Evidently, unless a woman takes an extraordinary initiative, she is not particularly encouraged to fulfill the prerequisites for securing admission to university courses in these traditionally male subjects. Achieving gender-equity in mathematics will hopefully result in women having opportunities to work in these areas as well.

Working Women

Although Papua New Guinean society is still male-dominated, change is taking place in the cities. A few women have managed to secure professional positions as managers, doctors, lawyers, secondary-school teachers, and university lecturers. However, most women who have found their way into the workforce perform jobs that do not require an advanced knowledge of mathematics. For instance, they are employed as typists, receptionists, domestic labourers, preschool or elementary teachers, nurses, and hospital orderlies. Primary-school teaching requires only elementary arithmetic and geometry. Secretarial work necessitates a basic knowledge of word processing and familiarity with fax machines, while jobs in banking and the travel industry require a modest level of experience with specialized computer packages. Nevertheless, while these jobs require some familiarity with technology, the related tasks are generally routine and require little in-depth understanding of the processes involved.

Learning Mathematics

Although mathematics is often called a universal language, there are a number of very real problems associated with students being forced to learn mathematics in a language other than their own. In Papua New Guinea, mathematics is most often taught in English which, for many students, is their second, or even third, language. As a result, in trying to understand a mathematical problem, students may have to resort to multiple translations. While English-speaking students can easily translate English words directly into their mathematical equivalent, those who speak English as a second language often find it necessary to translate words into their native language first. The situation is further complicated by the fact that many English concepts are quite unfamiliar to Papua New Guinean students and important aspects of a concept might conceivably be lost in the translation. English-language difficulties may also interfere with students' abilities to solve problems, for example because they find it hard to translate English statements into mathematical language. When the correct mathematical interpretations are provided, students are usually able to perform the mathematical steps required to arrive at a solution. But, without a firm understanding of English grammatical rules, students might have difficulty in constructing logical statements and writing analytical explanations.

Spatial skills, visual-processing abilities, the development of time concepts, and the use of comparatives are all areas in which student weaknesses have been identified (Priest, 1983; Sullivan, 1982). There are many time concepts extant in Papua New Guinean culture, and several clear and unambiguous ways of comparing objects and situations. Unfortunately, many students do not learn these concepts in their own language and context since they are sent to school at a time when they would normally be learning these ideas at home. In western countries, preschool teachers, as well as elementary teachers, devote considerable time and effort to having students use comparatives in a variety of contexts. Children also have access to construction toys such as 'LEGO' which help develop their spatial skills. Furthermore, teaching is based on language skills which the students already possess. In contrast, Papua New Guinean students must learn the language of instruction at the same time they are expected to learn the mathematical concepts. In view of this, it is hardly surprising that they would develop skill in time tasks and the use of comparatives more slowly than their western counterparts. Undoubtedly, this has a substantial impact on Papua New Guinean students' development of many higher-order mathematical concepts and procedures, such as in arithmetic, which depend on a solid understanding of these earlier concepts. A potential danger is that instructors may try to compensate for this by resorting to rote methods of teaching simply because the students have not developed a satisfactory comprehension of the prerequisite ideas. All of the foregoing might help to explain why, as has often been observed, Papua New Guinean students sit quietly in the classroom and are reluctant to participate in class discussions.

Conclusion

In this chapter I have identified a number of social and cultural issues in Papua New Guinea that contribute to disadvantaging girls in mathematics education. Other difficulties such as the language issues raised in the previous section, while common to all students regardless of gender, serve to compound the problem for female students. The question at hand is how can gender differences in mathematics education be addressed?

Some social reformers have attempted to use western strategies to deal with these problems without taking into account the deep cultural differences that exist and the centuries of deeply instilled traditional values that stand in the way of reform. For example, attempts to evoke change by using gimmicks such as mathematics medals, or competitions for girls, have resulted in complete failure. So have suggestions that the government create a separate examining board and a 'female curriculum' for girls. Instead, reforms that are grounded in an understanding that 'Rome was not built in a day' have been found to be more successful, especially in small, rural communities which resist sudden change. One such programme has worked within the cultural parameters of society and gently presented mathematics as an interesting and viable alternative to housewifery. The success of this particular approach is yet to be evaluated.

Many politicians and educators are optimistic that women are beginning to share equal educational opportunities. Rising tertiary enrolments and an increasing number of women in the workforce have been cited as examples. As yet, however, there are no indications that the situation is improving in the rural areas, but as more schools are built one expects progress to be the natural consequence. At the same time, there is a frightening phenomenon which overshadows the rural–urban debate. Law-enforcement officials have noted a rising level of violence against women. Even worse than the violence itself, is the fact that traditionalists have attributed it to the social reforms which have taken place since independence (Toft, 1985). Educational reforms have made it possible for women to have greater control of their lives as well as the right to demand an education. Traditionalists claim it has also resulted in the loss of male authority and that the violence is merely a way for men to protest against the disruption of their lives. The key to improving the quality of women's lives might well be extensive legal and social reform aimed at eliminating harmful traditional attitudes, rather than focusing on the symptoms, namely, women's low educational qualifications.

References

BACCHUS M.K., ERI, V. and MCNAMARA, V. (1985) *Report on Upper Secondary Education in Papua New Guinea*, Waigani, Department of Education.

DAVIDSON, M. and COOPER, C. (Eds) (1985) *Working Women: An International Survey*, New York, J. Wiley and Sons.

GUTHRIE, G. and SMITH, P. (Eds) (1980) *The Education of the Papua New Guinean Child: Conference Proceedings*, University of Papua New Guinea Publications.

KANAWI, J. (1986) *Marriage in Papua New Guinea*, Law Reform Commission of Papua New Guinea, Monograph No. 4., Port Moresby, Law Reform Commission of Papua New Guinea.

LANCY, D.E. (1983) *Cross Cultural Studies in Cognition and Mathematics*, New York, Academic Press.

LEAN, D.E. (1987) *Counting Systems of Papua New Guinea*, Vol. 1–17, Papua New Guinea University of Technology Publications.

OLIVER, J. (1985) 'Women students at the University of Papua New Guinea', *Council Report*, University of Papua New Guinea Publications.

PRIEST, M. (1983) 'A preparation for study in engineering and applied sciences', *Research in Mathematics Education in PNG.*

SULLIVAN, P. (1982) 'The development of time concepts and the use of comparatives among Papua New Guineans', *Research in Mathematics Education in PNG.*

TOFT, S. (1985) *Domestic Violence in Papua New Guinea*, Law Reform Commission of Papua New Guinea, Monograph No. 3., Port Moresby, Law Reform Commission of Papua New Guinea.

WORMALD, E. and CROSSLEY, A. (Eds) (1988) *Women and Mathematics Education in Papua New Guinea*, Waigani, University of Papua New Guinea Press.

Chapter 17

The French Experience: The Effects of De-Segregation

Françoise Delon

Introduction

France has a higher proportion of female university teachers and researchers in mathematics (24 per cent) than many other countries (Barsky *et al.*, 1987). This is often explained in part by the existence of the ENSs (Écoles Normales Supérieures), from which many of these teachers derive, and which until recently were single-sex institutions. The move to coeducation has had disastrous consequences for the number of women studying mathematics. This chapter explores some of the reasons for the ENSs' historical success in producing female mathematicians and raises some concern about recent developments.

The 'Écoles Normales Supérieure' System

The French system of higher education consists of state-funded universities with national standards, and 'Grandes Écoles'. The latter may be privately or publicly funded, possess varying entrance requirements, and offer a variety of programmes. In comparison with universities, the 'Grandes Écoles' prepare students for careers (for example, political, educational, administrative, commercial) and as such, have two complementary roles: the reproduction of the power elite and social advancement. They recruit by means of competitive examinations for which candidates study for two or three years after receiving their high-school diploma. Amongst the most prestigious of these 'Grandes Écoles', are the ENSs which have very high entrance standards and admission success rates varying between 5 and 8 per cent.

Students spend three or four years at the ENSs, often taking courses as well as examinations at a local university. While the original purpose of the ENSs was solely to produce secondary-school teachers, the range of career options now available is broader, including university teaching and research, and all forms of public-sector occupations especially in the senior administrative levels. In a variety of disciplines, including mathematics, many university teachers as well as researchers come from the ENSs. Consequently, ENS students enjoy a very high reputation in the universities and research institutions. Other advantages offered by the ENSs to their students are excellent conditions for learning and a network of information and professional relations to assist them in achieving their professional goals.

The Move to Coeducation

When the French State established a public, secular system of education in opposition to the existing religious one, it also created the ENSs with the role of training teachers for the new secondary schools. The oldest ENS was created at the rue d'Ulm in Paris and began functioning under Napoléon, at the start of the nineteenth century. Sèvres and Fontenay, both for women only, and Saint-Cloud (for men) were formed at the end of the nineteenth century. The ENSET (École Normale Supérieure de l'Enseignement Technique), created in 1912, was coeducational from its beginning. The ENSET, Saint-Cloud, and Fontenay schools attracted students from predominantly modest social backgrounds, while Ulm and Sèvres recruited students from high schools attended only by children of the bourgeoisie.

With the passing of the years, the policies of the different ENSs have been unified, although there are still noticeable differences in the social backgrounds, professional futures, and ideologies of their students. While there is no clear hierarchy among the other ENSs, Ulm, traditionally a men's school (with some nuances, see Hulin, 1994), is definitely the most prestigious. Several famous politicians, writers, and scientists are alumni of Ulm, and the institution prides itself on being the route to a brilliant career. It is considered the 'temple' of intellect and academia. The situation at Sèvres has evolved over time (Mayeur, 1994; Hulin, 1994), and has always been ambiguous. Despite sharing its entrance examinations and official statutes with Ulm since 1936, the faculty and alumni at Sèvres are not as powerfully established in the university community. Indeed in 1981, at the celebration of the centenary of Sèvres, while acknowledging that the ENS at Sèvres had accomplished part of its role in that women now have access to knowledge, several officials pointed out that women had yet to accede to power. Whether they were referring to the male-dominated world of politics, or to power in the intellectual world is not clear. However, what is certain is that they appeared to believe that women had no further need for the protective structure of the women's ENSs and, in fact, were perhaps even disadvantaged because of the poorer reputation of their schools. Soon after this event followed first the merger of Fontenay and Saint-Cloud, and then Sèvres and Ulm, without any serious prior analysis as to the possible consequences. The mergers did not prove to be favourable to women in any discipline (Ferrand, 1994). However, in no subject were the consequences as disastrous as in the area of mathematics, and at this point it would be appropriate to turn to a consideration of this phenomenon.

The Situation after Coeducation

The mathematics requirements of the entrance examinations for the individual ENSs were all identical. Furthermore, within each of the pairs, Sèvres/Ulm and Fontenay/Saint-Cloud, the written examinations were common, although assessment was done independently. In the past, each institution had a certain number of designated places for mathematics, about forty for the men's schools and about twenty for the women's schools (fewer females applied). In the eight years since the merger of Sèvres and Ulm, only twenty-one females have entered mathematics. According to Ferrand, Imbert, and Marry (1993) this amounts to a decline in the numbers of female mathematicians in these years of over 80 per cent.

Several factors contribute to this alarming statistic. Within the preparatory classes for the scientific 'Grandes Écoles', between 15 and 20 per cent of students in the most mathematically challenging sections are women (Barsky *et al.*, 1987). However, the corresponding figure stands at only 10 per cent in the most prestigious high schools which furnish the majority of ENS students. Here one finds the most dramatic form of the lingering estrangement that girls demonstrate where mathematics is concerned throughout their schooling. They often attribute their reluctance towards pursuing the subject to its 'abstraction' or its remoteness from human concerns (Mathématiques à Venir—Opération '50 lycées', 1988). It is important to realize, however, that in France mathematics is used as a gatekeeper to higher education and that an interest in mathematics conveys as much a concern for a professional future as a taste for the subject itself (Baudelot and Establet, 1992; Duru-Bellat, 1990).

Mathematics is, in fact, the key to the most prestigious professions. This phenomenon, which holds true throughout schooling, increases in impact in the classes which prepare students for the ENSs. The work required in this setting is more intense than at any other time in a student's education. It is highly structured and focused on its ultimate goal, that of admission to one of the schools which function as a springboard to employment at the top of the public sector and decision-making professions, a world that is still almost exclusively masculine. Many families, particularly those belonging to the upper social classes, expect their male offspring to have professional ambitions at an early age and plan their studies to this end. At the same time, they are often less concerned with the education of their female children. Prepared and supported by familial and societal expectations, boys tend to accept the constraints of preparatory classes more readily than girls. In addition, they also appear more comfortable with the competitive atmosphere found there.

This is the social context of the preparatory classes and one which functions to assist boys more than girls in concentrating and mobilizing their energies. There is, however, another aspect of the preparatory classes which should not be ignored. The oldest and most prestigious of them, and hence the ones which provide the greatest chance of success, have well-established rituals including their own special vocabulary, private jokes, initiation rites, and a tendency to form hierarchies, all of which might seem more in keeping with military school. There is a strong pressure on students to conform to the 'spirit' of the preparatory classes, and any resistance to this might be a source of difficulty for the student. Given their minority status in such an environment, women are often victimized and made the target of remarks of astonishing rudeness. (For an account of the 'initiation rites' in the preparatory classes and their discriminatory content, see Dupé, 1992.)

Thus, the preparatory classes do not constitute a particularly enjoyable experience, especially for women; solid motivation and a reasonable expectation of success are essential for those who commit themselves to attending them. Before the mergers, the women-only ENSs represented a reassuring alternative from several viewpoints: their existence allowed women to escape the severest forms of competition, and offered honourable and reasonable professional prospects. After all, were they not destined at the end of their ENS studies to obtain that most female of occupations — secondary school teaching? However, the professional expectations of the successful were changed by the new environment. In the ENSs there are greater expectations that students will pursue professions other than high-school teaching. The women-only ENSs made possible a smooth transition from traditional expectations to more career-oriented ones. This route is now more difficult for women. The female students of the mathematics preparatory classes know it well, for they apply in much smaller numbers to the coeducational ENS than they did to Sèvres. One may fear that the *de facto* closing of the ENSs to female mathematicians will push many female students out of the preparatory classes, with only those with early established professional goals remaining.

Women's 'Écoles Normales Supérieures'

Let's return again to an examination of the women's ENSs. Before coeducation, the ENSs had five to ten times the number of female mathematics students than they do today. Traditionally, women's ENSs have produced a large number of highly qualified teachers who enjoy a great intellectual prestige, and as such have provided generations of high-school girls with female academic role-models. These institutions have spawned many of the regrettably few scientifically renowned female figures, as well as facilitated the establishment of women in the university sphere (Coste-Roy, 1987; Barsky *et al.*, 1987). The women-only environment, which was both reassuring and stimulating, gave women an intellectual legitimacy which enabled them to enter mixed competitive environments such as professional recruitment examinations. Even if they met with less success than their male counterparts, the proportion of women who did achieve success was greater than would be expected given their actual numbers in the preparatory classes (Imbert, 1994). This world of women had the flavour of an utopia. Women's ENSs provided an image of intellectual women that was quite different from the stereotypical one, that being mostly single women enjoying books, test-tubes, courses, journeys, and meetings. Whether this new image had any validity is unimportant. But it was the myth, and a myth whose value lay in the provision of an image counter to that of the housebound wife without emotional autonomy, or the superwoman, seductive and successful in both career and family. The existence of this so-called 'other' exposed the contradictions and unrealistic dimensions of these misconceptions. It made it possible to imagine roles other than those provided in advance. It provided an environment where nobody would have thought of describing Emmy Noether as 'fat, rough, and loud'.[1] Myths aside, the women's ENSs were, for some of their students, crucial places for the acquisition of personal maturity and autonomy. It was at the ENS where visions of certain female figures became ingrained in the minds of students, helping them to establish themselves in the world as women.

Conclusion

The question now becomes what will happen next? If present-day figures concerning women mathematicians at the ENSs do not change, one may fear a significant decrease in the number of women in research and university positions in the future. The women's ENSs played a crucial role in promoting women in mathematics. They produced most of the pioneers and, up until recently, a sizeable portion of female researchers and university teachers, particularly those in senior positions. The demise of the single-sex ENSs came in the name of modernity, or even in the name of equity. The illusion consisted of thinking that coeducational institutions could be egalitarian in a world which is not.[2] The question, therefore, is how to remedy this? The idea of setting quotas is not popular in France. Furthermore, in the present context it is seen as contradicting the aura of objectivity attributed to mathematics, and to the principle of selection by competitive examination. Perhaps it will be possible to profit from European Community structures which permit affirmative action to offset historical discrimination. An analysis of current recruitment procedures may shed light on new possibilities, besides being healthy for the profession as a whole.

In the present system of preparatory classes and competitive examinations, the success of some is only possible because of the failure of others. The fact that many mathematicians are the product of such a process probably explains the lack of solidarity sometimes observed in the French mathematical community. In addition, the rigidity of such an environment favours the emergence of mathematicians with remarkably homogeneous intellectual and social behaviours. There may be much to gain by diversifying modes of training, criteria, age of recruitment, and ways of working. This approach might be preferable to the artificial promotion of particularly rigid careers or scientific profiles. Moreover, women may find themselves less torn between contradicting stereotypes if a more flexible system and ideology were adopted.

Notes

1 On a poster depicting mathematicians throughout history, Emmy Noether — the only woman appearing — is so described. This poster hangs in many mathematics departments and universities in North America.

2 There are no legal obstacles in France to employment equity. Theoretically, all professions are open to women. Three-fourths of women of working age do work, but, on the average, they are disadvantaged with respect to men in terms of salary, advancement, and employment.

References

BARSKY, D., CHALEYAT-MAUREL, M., CHRISTOL, G., GODBILLON, C., KLEIN, E., MÉRINDOL, J.-Y., MOEGLIN, C., OVAERT, J.-L., RAOULT, A., ROUSSIGNOL, M. and SIMON, J. (1987) 'Démographie des mathématiciens. Des mathématiques à l'industrie. La fuite des cervaux', *Bulletin de la Société Mathématique de France*, Supplément au tome 115, pp. 343–91.

BAUDELOT, C. and ESTABLET, R. (1992) *Allez les Filles*, Paris, Seuil.

COSTE-ROY, M.-F. (1987) 'La place des femmes dans les mathématiques: Problèmes actuels, perspectives d'avenir', *Bulletin de la Société Mathématique de France*, Supplément au tome 115, pp. 320–42 (in particular the contributions of J. Ferrand, N. Desolneux and D. Perrin).

DUPÉ, M.-O. (Ed) (1992) *Bizutages*, Condé-sur-Noireau, Arléa-Corlet-Panoramique n° 6.

DURU-BELLAT, M. (1990) *L'École des Filles*, Paris, l'Harmattan.

FERRAND, J. (1994) 'Une génération sacrifiée', *Sévriennes d'Hier et d'Aujourd'hui*, 148 (déc. 93–mars 94), pp. 15–17.

FERRAND, M., IMBERT, F. and MARRY, C. (1993) 'Normaliens, normaliennes scientifiques: l'excellence a-t-elle un sexe?', communication, colloque *Pour un nouveau bilan de la sociologie*, Paris, Institut National de la Recherche Pédagogique, 25–27 mai 1993.

HULIN, N. (1994) 'La section des sciences de l'École Normale Supérieure: Quelques jalons de son histoire', in SIRINELLI, J.-F. (Ed) *École Normale Supérieure. Le Livre du Bicentenaire*, Paris, Presses Universitaires de France, pp. 321–49.

IMBERT, F. (1994) 'Agrégations scientifiques et ENS au temps de la mixité', Unpublished manuscript.

Mathématiques À Venir — Opération '50 lycées' (1988) *Les Maths et Vous*, Strasbourg, Institut de Recherche sur l'Enseignement des Mathématiques.

MAYEUR, F. (1994) 'Sèvers', in SIRINELLI, J.-F. (Ed) *École Normale Supérieure. Le Livre du Bicentenaire*, Paris, Presses Universitaires de France, pp. 73–111.

Chapter 18

The Contribution of Girls-Only Schools to Mathematics and Science Education in Malawi

Pat Hiddleston

Introduction

Malawi is a small, land-locked, poverty-stricken country in Central Africa. Even by African standards Malawi is densely populated, with one million Mozambican refugees in addition to its own nine million people. The local economy is primarily based on agriculture, and as a result, rural settlement is the predominant pattern. In comparison with the western world, the country lacks technological advantages such as television and calculators: at school, for example, tables of logarithms are still quite commonly in use. The rate of participation in the educational system drops at each successive level. For instance, according to Ministry of Education and Culture (1990) statistics only 41.7 per cent of primary-school-aged children attend school, a figure which drops to only 3.4 per cent at the secondary level, and thereafter less than 1 per cent receive any form of tertiary education (Ministry of Education and Culture, 1990). Entry to secondary school is determined by means of a highly selective examination, the PSLE (Primary School Leaving Examination). A Ministry of Education affirmative action policy requires that one-third of all secondary-school places be reserved for girls, but in order to achieve this quota, many girls are admitted to school with lower PSLE scores than is characteristic of the boys.

Malawi's only university has five colleges, the largest of which, Chancellor College, incorporates faculties of science, humanities, sociology, law, and education. Admission to Chancellor College is based on grades in the final secondary-school examination known as the MCE (Malawi Certificate of Education). Girls account for only 23 per cent of the student population of Chancellor College, despite the fact that they are granted entry to university with lower scores than boys and comprise 34 per cent of the university student population as a whole (World Bank, 1988; Chancellor College, 1991).

Until recently, there has been very little research on women's issues in the developing world. In the few cases where such research has been undertaken, it has been conducted in isolation and has been difficult to publish. Yet over 75 per cent of the world's women live in developing countries. Since the needs of these women are often quite different from women in the developed world, research on developed nations cannot be assumed to apply universally. This chapter contributes to the growing body of literature on gender issues in the developing world by elaborating on several salient issues concerning students at Chancellor College. First, the majority of female students at Chancellor College attended girls-only schools and achieved higher MCE scores than their counterparts from coeducational facilities. Second, while female students entering Chancellor College do so with relatively low grades compared with males, especially in mathematics, they improve to a much greater extent and graduate with results that are, on average, higher than those of the male students.

Girls-Only Schools and University Selection

Developing countries have not been the target of much investigation concerning the relative merits of single-sex and coeducational schooling. However, independent research conducted both in Nigeria and in Thailand points to single-sex schooling as providing the most beneficial learning experience for girls (Lee and Lockhead, 1990; Jimenez and Lockhead, 1989). There are three categories of schools in Malawi: government, grant-aided, and private. The government schools may be either single-sex or coeducational, although the newer ones are mostly the latter. Many single-sex schools are grant-aided, and while they enjoy a certain degree of freedom from Ministry of Education regulations, they are still compelled to adopt the same selection procedures and charge the same fees as government schools. Therefore, in most respects these two types of educational facility are similar. There are so few private schools that they may be disregarded with little impact on the analysis.

Table 18.1 illustrates the performance of girls in the MCE examination for the years 1987 to 1991. It shows that the percentage of MCE passes in each year was consistently higher in the girls-only schools. At the same time, the coeducational schools were seeing a much greater increase in the number of girls taking the examination, undoubtedly due to the newer government schools which were predominantly coeducational. The Ministry of Education insists that school selection is independent of the academic standing of the students and that there is no bias towards placing the best girls in single-sex schools.

If there is any validity in the Ministry of Education's claim that they do not place the brighter students in single-sex schools, one would expect the numbers of girls admitted to Chancellor College from both types of institution to be roughly in proportion to the numbers who pass the MCE. Table 18.2, however, shows that from 1987–1991, the number of girls entering Chancellor College from girls-only schools was substantially higher. This pattern persists when one restricts the data to those entering the two science-based programmes, or indeed any undergraduate programme, and it has not changed as the intake from

Table 18.1: MCE performance of girls from coeducational and girl-only schools for 1987–1991

	Coeducational			Girls-only		
	Sat	**Passed**	**Percentage**	**Sat**	**Passed**	**Percentage**
1987	510	253	49.6	901	643	71.4
1988	583	305	52.3	1067	722	67.7
1989	1026	471	45.9	1073	599	55.8
1990	1106	452	40.9	1294	760	58.7
1991	1068	414	38.8	1260	745	59.1

Source: Chancellor College (1991) *Statistics*

Table 18.2: Intake of girls to Chancellor College, as a percentage of total intake, for all degree courses by type of school

	Total intake	Girls as a percentage of total intake	Percentage girls from coed schools	Percentage girls from girls-only schools	Percentge girls from other schools
1986–87	227	17	1	13	3
1987–88	293	14	3	10	1
1988–89	278	24	3	18	3
1989–90	329	18	3	13	2
1990–91	379	23	4	17	2
1991–92	409	19	4	12	3

Source: Chancellor College (1991) *Statistics*

coeducational schools has increased. All in all, the situation in Malawi education appears to be consistent with that in both Nigeria and Thailand, that is, students from female-only schools meet with greater academic success than their coeducational sisters.

Gender Comparisons of Performance

At all stages of their school careers, girls in Malawi are outperformed by their male counterparts in mathematics. In fact, PSLE scores for the years 1989 to 1991 indicate that primary-school boys achieved better results than girls in all subject areas (Bradbury, 1991). Kadzamira (1987) has documented a similar trend for high-school students on the MCE in the period 1982 to 1986. As previously discussed, affirmative action policies result in the admission to Chancellor College of women with qualifications that are comparatively lower than those of men. Moreover, the situation is particularly acute in the area of mathematics. However, when one examines performance in the first-year mathematics course, one notes a substantial improvement in female performance over the year relative to males.

Table 18.3 documents this trend for the years 1987 to 1991. It shows women's

Table 18.3: First-year mathematics grades at Chancellor College from 1987 to 1991

	Total	Total female	MCE (all)	MCE Maths	Test 1 Nov	Test 2 Feb	Test 3 Mar	Final Exam June
87/88	223	25	50.2	45.5	49.5	49.4	49.2	45.2
88/89	175	29	47.9	46.1	49.0	50.1	51.6	51.8
89/90	206	32	46.3	45.8	43.2	47.3	49.1	50.2
90/91	258	55	45.5	43.4	45.2	46.2	47.0	47.7
Average			47.0	44.9	46.3	47.8	48.8	48.7

Source: Chancellor College (1991) *Statistics*

Table 18.4: Second-year mathematics grades at Chancellor College from 1988 to 1992

			MCE		Year One				Year Two		
	Total	F	All	Math	T1	T2	T3	Ex	T1	T2	Ex
88/89	136	11	52.4	47.5	52.3	54.1	51.0	50.5	48.8	52.3	49.6
89/90	100	16	47.4	46.9	48.3	48.5	51.1	51.7	–	–	54.4
90/91	115	14	49.8	47.7	43.5	50.4	51.6	52.6	51.3	52.6	52.0
91/92	141	26	44.1	40.6	45.6	46.1	45.5	47.6	45.0	47.9	49.8
Average			47.4	44.7	46.9	48.9	49.0	50.1	47.5	50.1	51.3

Source: Chancellor College (1991) *Statistics*

scores, standardized to a mean of 50 for the entire cohort, for three term tests as well as for the final examination. Taking an average of all the years shown, the grade improvement throughout the course is quite apparent. The fact that all students were taught and evaluated by the same (male) lecturer, ensured the necessary level of control.

Table 18.4 documents continued improvement into the second year, with the data for first-year students adjusted to include only those who did, indeed, advance to take the second-year mathematics course. It is interesting to note that the female average for all years shown of 51.3 on the second-year final exam exceeds the corresponding figure of 50 for the whole group. In essence, this means that on completion of their second mathematics course, female students, on average, achieve at least as well as, if not better than, their male counterparts. (Again, the conditions were controlled in the same manner as before.)

This trend of improved performance is not confined to the first two years of mathematics, but extends to the sciences as well as into more advanced mathematics courses. Table 18.5 shows average female-standard (Z) scores for the years 1981 to 1990. Seven subject areas are included, biology, chemistry, computer science, earth science, mathematics, physics, and statistics. Computer science is not offered in year one, while statistics is offered only in years three and four. In

Table 18.5: Z Scores for end-of-year grades in the Science Faculty, Chancellor College, by year of study

	Biology	Chemistry	Computer Science	Earth Science	Maths	Physics	Stats
1	0.38	0.27		0.24	–0.36	–0.54	
2	0.30	0.00	0.26	–0.40	0.05	–0.45	
3	0.27 0.06 0.96	0.08 0.17 0.32	0.39	0.68 1.19	0.01 0.93	–0.10 0.15	–0.13
4	0.22 0.34 0.27	0.43 0.05 0.36	0.52 0.11	1.35 0.36	0.94 0.45	1.81 0.82	0.02

Source: Chancellor College (1991) *Statistics*

addition, as is illustrated in the table, certain disciplines have more than one course at some levels.

The data presented in the table for year one shows that all subject areas, apart from earth science, have negative Z scores. This means that female students generally perform below average in the indicated courses. However, as one progresses through each of the years, the overall trend is for the scores to gradually increase until they are all positive. Thus, by year four, female students are achieving above average in all documented subjects. These findings agree with those of Maritim (1985), whose study of Kenyan high-schools concluded that girls with lower high-school entry grades appear to outperform boys in their final mathematics examinations. These results contrast, however, with the majority of research findings in the developed world. In developed countries, male academic performance tends to surpass that of females in early adolescence, a trend which continues well into later years (Burton, 1990).

Several factors may combine to explain women's apparent success at Chancellor College. First, while Malawian school teachers expect girls to have lower grades than boys in mathematics and science, the staff at Chancellor College do not share this opinion. Indeed, they have no preconceived expectations. Second, women who enter the college experience a sense of personal achievement on having done so, and claim no longer to doubt their abilities. Thus, there is a shift from a self-fulfilling expectation of failure, to the anticipation of success. Third, women at Chancellor College have a reputation for working harder than men, possibly, as has been suggested, because they feel obliged to justify being admitted in the first place. Finally, Chancellor College is a residential institution. As a result, women are able to concentrate on their studies without having to contend with domestic chores. In contrast, school girls are often forced to be absent from school for domestic reasons, an expectation that does not hold for boys.

Conclusion

Females in Malawi who attend girls-only schools achieve better results on the final school examination than those who attend coeducational institutions. This is true even though they are selected for these schools by a process which, the Ministry of Education claims, has no bias towards placing the more academically successful girls in single-sex learning environments. Far more women are admitted to Chancellor College from girls-only schools, than would be expected from the proportion qualifying. While they enter the university with lower qualifications than men, women's results in the mathematics and sciences improve substantially. By the time they graduate, females outperform their male counterparts in these subject areas.

References

BRADBURY, R. (1990) *Differences Between the Results of Boys and Girls in Subjects of the Primary School Leaving Examination*, Malawi Examination Board, RTC 14/90.

BURTON, L. (1990) *Gender and Mathematics: An International Perspective*, London, Cassell.

CHANCELLOR COLLEGE (1991) *Statistics*.

JIMENEZ, E. and LOCKHEAD, M.E. (1989) *The Relative Effectiveness of Single Sex Schools in Thailand*, Educational and Policy Analysis.

KADZAMIRA, M. (1987) *Sex Differences in the Performance of Candidates in the Malawi Certificate of Education in Mathematics and Science Subjects*, Malawi Examination Board, RTC 21/87.

LEE, V.E. and LOCKHEAD, M.E. (1990) 'The effects of single sex schooling on achievement and attitudes in Nigeria', *Comparative Education Review*, 34, 2, pp. 209–31.

MARITIM, E.E. (1985) 'The dependence of "O" and "A" level results on the sex of the examinees', *Kenya Journal of Education*, 2, pp. 21–46.

MINISTRY OF EDUCATION AND CULTURE (1990) *Education Statistics*, Lilongwe, Malawi Government.

WORLD BANK (1988) *Education in Sub-Sahara Africa. Policies for Adjustment, Revitalisation, and Expansion*, Washington, DC, World Bank.

Part 3

Feminist Pedagogy in Mathematics Education

The fourth phase of gender reform of mathematics education, described in the introduction asks the question: What would mathematics and mathematics education look like if women were central to its development? We have divided the chapters in Part 3 into two categories according to whether they emphasize changing mathematics pedagogy or changing the discipline. Although these depart from different standpoints, they come to similar conclusions and lead us to speculate as to the nature of an inclusive mathematics education.

The influence of feminism on approaches for achieving equity in mathematics education has perhaps been most evident in the area of pedagogy. Some of the chapters in this section provide a theoretical description of a pedagogy based on an understanding of women's ways of knowing, others give practical suggestions for implementing a feminist mathematics pedagogy. Whatever their focus, all share the desire to see a fundamental change in the distribution of power in the classroom and as a consequence in the organization of the discipline of mathematics as a whole.

In the first chapter, Roberta Mura defines three strands within contemporary feminism and demonstrates how all approaches to gender reform of mathematics education are, to different extents, influenced by some form of feminism. The chapter by Joanne Rossi Becker describes stages in women's ways of knowing, the connected teaching that grows out of an understanding of this model, and examples of how to apply this theory to mathematics. The practical implementation of a feminist pedagogy is the focus of the chapter by Pat Rogers. Drawing on Gilligan's theory of gender-related differences in moral development, she describes how she abandoned the authoritative lecture format which silences women and began developing a creative-intuitive pedagogy that has been particularly successful with her female undergraduates. Sue Willis, based on a feminist critique of previous proposals to change the gender imbalance in school mathematics, develops proposals for a different mathematics curriculum, one which is gender-inclusive and socially critical. This implies, among other things, non-sexist curriculum content, the use of a variety of intuitive and exploratory approaches to mathematics, connecting mathematics to reality, and the explicit confronting in the classroom of gender issues in the wider community. The focus of Joanna Higgins' chapter is the use made by teachers of games as independent activities in the teaching and learning of mathematics at the elementary school level. She argues that the classroom experiences of girls differ from those of boys in ways that disadvantage the girls and so the structures of learning experiences if left unexamined tend to benefit boys more than girls.

The chapters of Leone Burton, Betty Johnston and Marjolijn Witte, included in the second group, all deal with the need to change the discipline of mathematics. They ask whether there exists a female mathematics and whether it would be different from mathematics as developed thus far. Leone Burton, referring to the extensive literature on gender and the nature of science, asks similar philosophical, pedagogical and epistemological questions about the discipline of mathematics. Betty Johnston reflects instead on the interaction between mathematics and society, in particular on how the quantification of society constructs us and how we, in turn, construct or resist it. Through memory work, she explores using personal experience as the basis of knowledge and examines the interface between the everyday world and wider social structures. One particular thread that emerges is the experience of mathematics as regulation by others, a thread that is also picked up by Marjolijn Witte. Distinguishing several different strategies for intervening to prevent girls from dropping out of mathematics education, Witte explains how the dominant language in mathematics constrains the actions of learners and disadvantages girls. Within this language conception of mathematics, she discriminates two basic conceptions of mathematics, one of which she claims holds promise for girls.

Chapter 19

Feminism and Strategies for Redressing Gender Imbalance in Mathematics

Roberta Mura

A variety of theoretical perspectives has been advanced for addressing the issue of gender imbalance in the teaching and learning of mathematics. For instance, in Kaiser-Messmer and Rogers (1994), four examples are given: 'The "intervention perspective" which locates the problem in the student; the "segregation perspective" in which the interaction between girls and boys is often the primary focus; the "discipline perspective" in which the nature of mathematics itself is problematized; and the "feminist perspective" in which a critique of the gendered nature of teaching and learning mathematics leads to an examination of the use of power and authority in the mathematics community' (Ibid., p. 304). For me, this raises some questions. Why did the authors label only one of these four perspectives feminist? Don't all approaches which seek to address the imbalance correspond to some kind of feminist analysis and practice? Indeed, how *does* feminism relate to work on women and mathematics?

Although the word 'feminism' usually occurs in the singular, there are in fact a variety of feminist theories. In this chapter, I consider the possible links between different kinds of feminisms and the different approaches to the issue of gender imbalance in mathematics education described above and illustrated by four selected positions, namely, those of Sheila Tobias (1978), Charlene and James Morrow, Pat Rogers and Sue Willis (see Chapters 2, 21 and 22, respectively, this volume). I begin with a brief discussion of feminist theory.

Trends within Feminism

It is likely that feminism means something different to every feminist, and something else again to those who oppose it. All feminists probably would agree that women suffer certain disadvantages in comparison with men. Beyond that, they might analyse the situation differently and favour different strategies for changing it. In this section, I draw on a typology proposed by Descarries-Bélanger and Roy (1991) to identify three main trends within contemporary feminism:

'feminism of equality', 'radical feminism', and 'feminism of difference' (no hierarchy is implied by the order used here).[1]

Feminism of equality is action-oriented; it demands legal and actual equality between women and men, and it identifies the sexual division of labour as the main source of women's oppression. It denounces, first and foremost, the discriminatory conditions women experience in the spheres of education and work, and advocates programmes that would ensure equal access for women to all social, political and economic resources. Together with political action, the socialization and education of girls are seen as the most effective tools for change. Of the three feminist currents discussed here, this is the one that comes closest to the typical definition of feminism found in English and French dictionaries. Feminists of equality might extend their criticism of sexual discrimination to all other kinds of discrimination, however they do not necessarily challenge social institutions, they ask 'only' that women occupy 50 per cent of all political offices as well as all military, religious, academic, managerial, judicial, and other offices. Some consider feminism of equality reformist, while others think that true equality between the sexes in all social, political and economic spheres would be tantamount to a complete revolution.

Radical feminists refuse to define themselves in comparison with men.[2] They identify patriarchy as a social, political and economic system that oppresses and exploits women individually and collectively, sexually and economically. Radical feminists have put more energy than feminists of equality into theoretical productions and their positions are also more diversified. Descarries-Bélanger and Roy (1991) distinguish three tendencies within the radical current, namely, materialist feminism, 'woman-centred' feminism, and lesbian feminism. I won't dwell on these distinctions here. Radical feminists see patriarchy as characterized by hierarchical relations between the sexes, maintained through social institutions like marriage and heterosexuality, through laws, religion and the educational system, as well as through violence or the threat of violence. Radical feminism refuses the separation between domestic and public life and questions the androcentric perspective that pervades all aspects of patriarchal cultures. Strategically, priority is given to ending all types of violence against women and to actions aimed at the disappearance of the institutions responsible for their oppression and exploitation. Radical feminists might extend their criticism of the oppression of women to all other kinds of oppression.

Unlike feminists of equality and radical feminists, feminists of difference do not wish to eliminate gender distinctions. On the contrary, they insist on the recognition of difference, they see women as possessors of specific knowledge, culture and experiences and the feminine as an affirmation of life. They assert that women have their own ethics, their own way of knowing and their own language. Their project involves reclaiming and revaluing all things feminine. Influenced by psychoanalytical theory, they want to discover and liberate a truly feminine sexuality that has been repressed by patriarchal culture. Since women's specificity consists principally in their childbearing and nurturing capacities, motherhood is at the heart of feminism of difference. Feminists of difference demand a better deal for wives and mothers, and are fierce opponents of the new reproductive technologies which they see as threatening to take away from women their control of motherhood and undermining their very definition. Feminists of difference promote the idea of the complementarity of the sexes and ask that women be allowed to engage in activities separate and different from men, but that these activities be equally valued and rewarded.

As with all classifications, this one distorts reality: individual feminists may hold ideas and participate in actions typical of two or all three trends. However, I believe that these categories reflect real differences within the women's movement. Sometimes, they are just a matter of focus and priorities, but sometimes they correspond to serious disagreements, where the proponents of opposing theories would deny each other the right to call themselves feminists.

Redressing the Gender Imbalance in Mathematics

What connections can be found between the three feminist currents described above and the perspectives on gender imbalance in mathematics discussed earlier? Clearly, redressing the gender imbalance in the teaching and learning of mathematics is part of a more global project of achieving educational and occupational equity between the sexes and is in itself a typical feminism-of-equality issue. Therefore, this current underlies all approaches seeking to redress imbalances. The intervention perspective, though, is especially characteristic of this brand of feminism, because of its emphasis on increasing the participation of women in mathematics and on programmes aimed at resocializing girls, that is, correcting, or compensating for, a deficient socialization. The early work of Tobias (1978) illustrates this perspective well, by espousing the goal of getting girls to choose mathematics and to persist in it, by focusing on an attitude of the individual student, namely mathematics anxiety, as the source of her mathematics avoidance, by assuming that this attitude is learned and therefore changeable, and by acting to help students unlearn it.

In contrast, the segregation perspective, is reminiscent of feminism of difference, in that it asserts that boys and girls have different ways of learning and that they are better taught separately, using methods and/or curricula appropriate to each. Whatever the origin of gender differences in learning styles, proponents of the segregation perspective consider it a given, either unchangeable or worth preserving. Support for this view is found in research like that of Belenky *et al.* (1986).

The Morrows (see Chapter 2, this volume) recommend the segregation option on other grounds. They focus on the interaction between girls and boys. In a mixed group, they argue, boys always end up getting a greater share of the available resources than girls. Although the Morrows attribute this gender difference in behaviour to socialization, they advocate segregation as a way of freeing girls from the oppressive situation that prevails in coeducational environments, for changes in socialization, much as they are desirable, would consume more time and energy than teachers and students usually can afford during mathematics classes. This argument proceeds from a radical feminist standpoint which analyses classroom interactions in terms of oppression and dominance and aims at girls' autonomous development without reference to boys.[3] It claims neither equality nor difference.

The discipline perspective is described above as the view that it is the discipline of mathematics itself that leads to the gender imbalance. Both radical feminism and feminism of difference may be relevant to this view. The former has launched an ambitious project of examining all existing knowledge for evidence of androcentrism. In spite of the apparent difficulty of the task, one may want to submit mathematics and the physical sciences to the same kind of feminist criticism that is being applied to other disciplines. Feminists of difference, furthermore, may entertain the hypothesis that mathematics, as it presently exists, is the product of a male way of thinking and may speculate about the possibility of a female mathematics. These ideas, however, are regarded with suspicion and hostility by mathematicians subscribing to feminism of equality, some of whom fear that feminist critiques of science may discourage women from entering the field.[4] I believe this fear to be unwarranted since several disciplines, such as biology, psychology, and literature, at least in Canada, have higher female participation rates than mathematics, despite being the target of much harsher feminist critiques.

Willis (see Chapter 22, this volume) espouses the discipline perspective, but in doing so she focuses on school mathematics and on the image of mathematics rather than on the discipline itself. Of course, one could argue that school mathematics is part of mathematics, however Burton (1987) points out that the features of school mathematics singled out as potentially responsible for the gender imbalance are precisely those that distinguish it from the discipline as actually practised.

The feminist (the fourth) perspective, discussed earlier places responsibility for the gender imbalance on the teaching of mathematics and calls for improved teaching methods that will benefit male and female students alike. All three feminist trends, not to mention the whole mathematics education community, have an interest in this perspective. Feminists of equality demand non-sexist teaching, that is, fair treatment of female and male students. Feminists of difference specify that fairness must take into account difference, and that equal treatment does not necessarily constitute equity. They insist on a pedagogy consonant with women's psychology, a pedagogy sensitive to students' emotions and based on collaboration rather than competition. Radical feminists favour a pedagogy that addresses the subject of women's oppression and liberation explicitly in the classroom. In line with radical thinking, this pedagogy seeks to minimize hierarchical relations between teacher and students, empowering female students, and forcing male students to give up their dominance in the classroom. As mentioned earlier, this type of feminist analysis of what goes on in the mathematics classroom naturally 'leads to an examination of the use of power and authority in the mathematics community' (Kaiser-Messmer and Rogers, 1994, p. 304), and beyond.

Both Rogers and Willis (Chapters 21 and 22, this volume) advocate a radical feminist pedagogical perspective and they both extend it to a more inclusive pedagogy of liberation, drawing on Freire's ideas (1971), exposing the role played by mathematics in constructing and maintaining privilege, be it gender, class or race privilege (see also Frankenstein, 1983, 1991, and Rogers, 1990).

It seems to me that the theoretical perspectives for redressing the gender imbalance in mathematics that have been outlined above could more precisely be called strategies, and that three emerge from the approaches discussed: the intervention strategy, the segregation strategy and the teaching strategy. The last strategy corresponds to the perspective identified as 'feminist' by Kaiser-Messmer and Rogers and incorporates aspects of the discipline perspective in that it concerns both pedagogy and curriculum. I have argued that each of these strategies can be linked to at least one feminist current. That all three can lay claim to feminism follows from their common intent to improve the situation of women in mathematics.

The claim to feminism of all three strategies does not preclude criticism of any of them for being less than effective, or even for being counter-productive. Such a debate parallels the one about the different tendencies within feminism and their potential for improving the situation of women in society. For instance, one could find fault with the intervention strategy for laying the responsibility for redressing the balance at the feet of girls and women (that is, for assuming that it is they who have to change) and for failing to criticize mathematics, mathematics education, or the educational system as presently configured. Like the feminism of equality underlying it, the intervention strategy may seem to encourage women to reach for half the pie without stopping to consider whether the pie is worth eating.

The segregation and the teaching strategies, depending on whether they draw on feminism of difference or on radical feminism, may be criticized for either reinforcing gender stereotypes (women and men think and learn differently)[5], or antagonizing men and exacerbating the tension between the sexes. Feminists of difference constantly risk playing into the hands of sexist interests. In the case of mathematics, any reference to gender differences may be interpreted or misinterpreted as supporting the thesis of male superiority in this field. Historically, sex segregation has not always been implemented with feminist goals in mind, and separate education has often meant inferior education for girls.

I would like to conclude this section with a few reflections on the general goal of redressing the gender imbalance in mathematics. First, there is the possibility that some feminists might not subscribe to this goal. The argument has been advanced that certain institutions (for example, the army or the church) are so fundamentally antithetical to women that it is preferable to stay away from them. So far, to my knowledge, no feminist has publicly advocated such an extreme position concerning mathematics; even the most severe critics of this discipline do not advise women to turn their backs on it. However, the theoretical possibility of rejecting mathematics, from either a radical feminist or feminism-of-difference standpoint, remains open.

Second, although this is rarely discussed, underlying most efforts to redress imbalance is the tacit assumption that the proper balance is roughly equal numbers of women at all education and occupational levels, up to the highest rank of professorship in the most prestigious universities. Even feminists of difference, while insisting on the qualitative aspects of the balance, have not questioned this quantitative target. To achieve this ideal proportion, in many countries, efforts to increase the percentage of women in mathematics have focused on increasing their number. However, there is another way in which the percentage of women in any field may increase, and that is when the number of men decreases!

For groups other than feminists, increasing the number, rather than the percentage, of women in mathematics may be a goal in itself, and not necessarily a means of achieving gender balance. Governments, for instance, may be interested in increasing women's participation in mathematics, science and technology merely because they want to increase the total population involved in scientific and technological production. Likewise, mathematics departments may seek to attract more female students as a means of remedying declining enrolments. Although feminists may take advantage of such situations, it is important to keep in mind the distinction between the goal of increasing women's participation and the goal of redressing gender imbalance. How would we react to the idea of the percentage of women in mathematics climbing to 70 or 90 per cent? And, would we then be happy to see women fill positions that men no longer wanted?

Conclusion

Gender imbalance, in mathematics as elsewhere, is a political issue. I have argued that strategies for redressing the imbalance rest on a variety of feminist philosophies that differ considerably from each other.[6] Making these underlying philosophies explicit will increase our ability to evaluate the possible advantages and dangers of each approach. Furthermore, it will show that we do not all share the same attitudes and ideals, and it will prevent the frustration caused by the unrealistic expectation that we do.[7] Because of the relevance of philosophical and political perspectives to our work, we probably will not be able to agree on what the best strategy is for redressing gender imbalance. This is not peculiar to our field. To choose another example close to our experience, mathematics teaching too is shaped by diverse philosophies of mathematics and education, even though they may remain subconscious, and for this reason it is unlikely that we will ever reach a consensus on what is the best way to teach mathematics. However, being aware of our often implicit perspectives will contribute to a better understanding of the motivations, aims and effects of the various options, and the extent to which they are consonant with our beliefs and values.

Notes

1 Several other frameworks have been employed for classifying feminist trends. Some rely on political or philosophical schools of thought or social movements that exist independently of feminism (for example, black, liberal, socialist, Marxist, existentialist, structuralist, post-modernist). Offen (1988) distinguishes between individualist and relational feminism, others refer to successive waves or generations of feminism (not necessarily defined by chronology). I have chosen Descarries-Bélanger and Roy's typology because it is more intrinsic to feminist thought and less open to value judgments.

2 The word 'radical' is used here, in the sense of 'root', to describe someone who wants to go to the root of a problem, to concentrate on its most fundamental aspects.

3 A similar standpoint is taken by Spender (1982). Spender shows how, in mixed classes, boys learn to dominate girls and girls to submit. For girls, she argues, mixed-sex education turns into indoctrination and practice in the art of subordination (Ibid., pp. 118–20). See also Chapter 23, this volume.

4 I have described at length elsewhere both the project of a feminist critique of mathematics and science (Mura, 1991, 1987) and mathematicians' reactions to it (Mura, 1992–1993).

5 Hanna (1994) articulates this kind of criticism.

6 Research on gender differences, of course, is another matter. It too must rest on some philosophy of gender relations, but not necessarily on a feminist one.

7 Lee (1992) provides a good illustration of contrasting attitudes towards gender issues in mathematics.

References

BELENKY, M.F., CLINCHY, B.M., GOLDBERG, N.R. and TARULE, J.M. (1986) *Women's Ways of Knowing: The Development of Self, Voice, and Mind*, New York, Basic Books.

BURTON, L. (1987) 'Women and mathematics: Is there an intersection?', *International Organization of Women and Mathematics Education Newsletter*, 3, 1, pp. 4–7.

DESCARRIES-BÉLANGER, F. and ROY, S. (1991) *The Women's Movement and its Currents of Thought: A Typological Essay*, The CRIAW Papers, 26, Ottawa, Canadian Research Institute for the Advancement of Women.

FRANKENSTEIN, M. (1983) 'Teaching radical math: Taking the numb out of numbers', *Science for the People*, 15, 1, pp. 2–17.

FRANKENSTEIN, M. (1991) 'Criticalmathematics education: Towards a definition', in QUIGLEY, M. (Ed) *Proceedings of the 1991 Annual Meeting of the Canadian Mathematics Education Study Group*, St. John's, Memorial University of Newfoundland, pp. 109–21.

FREIRE, P. (1971) *Pedagogy of the Oppressed*, New York, Seaview.

HANNA, G. (1994) 'Should girls and boys be taught differently?', in BIEHLER, R., SCHOLZ, R.W., STRÄSSER, R. and WINKELMANN, B. (Eds) *Didactics of Mathematics as a Scientific Discipline*, Dordrecht, Kluwer Academic Publishers, pp. 303–14.

KAISER-MESSMER, G. and ROGERS, P. (1994) 'Gender and mathematics education', in GAULIN, C., HODGSON, B.R., WHEELER, D.H. and EGSGARD, J.C. (Eds) *Proceedings of the 7th International Congress on Mathematical Education*, Sainte-Foy, Les Presses de

l'Université Laval, pp. 304–9.
Lee, L. (1992) 'Gender fictions', *For the Learning of Mathematics*, 12, 1, pp. 28–37.
Mura, R. (1987) 'Feminist views of mathematics', *Association for Women in Mathematics Newsletter*, 17, 4, pp. 5–10. Reprinted in *International Organization of Women and Mathematics Education Newsletter*, 3, 2, pp. 6–11.
Mura, R. (1991) *Searching for Subjectivity in the World of the Sciences: Feminist Viewpoints*, The CRIAW Papers, 25, Ottawa, Canadian Research Institute for the Advancement of Women.
Mura, R. (1992–1993) 'Les critiques féministes de la science: Une menace aux femmes et à la science? Analyse de deux réactions du milieu mathématique', *Atlantis*, 18, 1 and 2, pp. 3–24.
Offen, K. (1988) 'Defining feminism: A comparative historical approach', *Signs: Journal of Women in Culture and Society*, 14, 1, pp. 119–57.
Rogers, P. (1990) 'Thoughts on power and pedagogy', in Burton, L. (Ed) *Gender and Mathematics: An International Perspective*, London, Cassell, pp. 38–46.
Spender, D. (1982) *Invisible Women: The Schooling Scandal*, London, Writers and Readers.
Tobias, S. (1978) *Overcoming Math Anxiety*, New York, W.W. Norton and Co.

Chapter 20

Women's Ways of Knowing in Mathematics

Joanne Rossi Becker

That roars which lies on the other side of silence.

(George Eliot, 1871–2)

Introduction

In this chapter, I discuss the ideas in a book that I think has major implications for how we can encourage girls and women to pursue mathematics and mathematics-related careers.[1] *Women's Ways of Knowing: The Development of Self, Voice, and Mind,* by Mary Field Belenky, Blythe McViker Clinchy, Nancy Rule Goldberger, and Jill Mattuck Tarule (1986), explores how women come to know. After briefly summarizing the major ideas in the book, I will discuss implications for the teaching of mathematics, and raise issues which I think would benefit from further research.

Background

Feminist theory is usually developed in one of two ways. Individuals, either address, in the initial stages of their work, how a particular issue affects or is affected by women and their concerns, or they re-examine the work of others to see if their conclusions fit with women's experiences. All too often, conclusions based on work with male subjects are presented as though they are indicative of all persons.

It is the second approach to developing feminist theory that has received the most attention. These re-examinations are necessary because so much research has been based on all-male samples and thus the theories generated often result in women becoming a disadvantaged group. Probably the best-known effort to critique a widely accepted theory and assess its validity for women is Carol Gilligan's work (Gilligan, 1982). Gilligan examined Kohlberg's stages of moral development, which he based on an all-male sample, and asked whether the hierarchical stages he proposed were truly reflective of the way in which moral development occurs in women. She found they were not. In terms of Kohlberg's hierarchy, women seldom reach the highest levels of moral development. Gilligan, however, found that moral development in women followed a different path and was based on different values. In a profound way, Gilligan made the case that women were speaking in 'a different voice'. It was not that the values and world view of women were better or worse than those of men, it was just that their systems were different. Gilligan called for an honouring of both value-systems, as well as for both voices to be heard, thereby freeing everyone to look at how and what they value, and decide which way is right for them individually.

It has become acceptable for women to proclaim their 'different voice' with regard to moral issues. It is far more dangerous, however, to broach the hypothesis that women have a 'different voice' with regard to the cognitive. The questions are: Is the way we, all of us, are supposed to know things also determined by how some males came to know things? Is there a male model for knowing that, in its very formation, excluded women, denied their truths, and made them doubt their intellectual competence? In *Women's Ways of Knowing*, Belenky and her colleagues explore the ways in which women know, and also how these ways of knowing differ from those of men. In particular, they present a re-examination of William Perry's model of intellectual development (Perry, 1970).

Let me address the reservation that some of you may have about describing a particular way of thinking as 'women's way'. When I refer to women or girls in this chapter, as when the book refers to women in *Women's Ways of Knowing*, I am not automatically referring to all females. Here I am generalizing, not as mathematicians do, meaning every woman, but as social scientists do, meaning most women. The word 'women' is used to refer to all those individuals who think, come to know, or react in a fashion that is common to the majority of women. These individuals may be females or males. Also, the use of the word 'women' to describe the way some people think does not preclude the possibility that some women do not think in this way.

There is, of course, the danger that acknowledging women's different ways of knowing will serve to reinforce stereotypes that demean women's capabilities (see also Mura, Chapter 19 this volume). We have to make the case that 'different' does not mean one way of thinking is better than another. On the other hand, research can help provide evidence to support or refute the possibility of a women's way of knowing in mathematics and, if it is supported, demonstrate how this can help us understand and improve women's participation in mathematics.

Women's Ways of Knowing

The authors of *Women's Ways of Knowing* spent years interviewing 135 women of various ages and socio-cultural backgrounds. These women were African-American, European American, and Hispanic, and were identified in formal sites of education such as prestigious women's colleges, coeducational colleges, urban

Table 20.1: Stages of Knowing

Stage of Knowing	Statement
Silence Accepts authority's verdict as to what is true.	An inner voice expresses awareness that teachers think base angles are equal.
Received knowing Learns by listening; returns words of authority. Speaker is not source of knowledge.	'I know that base angles are equal because my teacher says so.'
Subjective knowing Inner voice says 'I only know what I feel in my gut.' Assumes there are right answers. Male version: 'I have a right to my opinion.' Female version: 'It is just my opinion.'	'I know that base angles are equal. Just look at them; they're equal.'
Procedural knowing Voice of reason; begins to evaluate validity of argument.	
Separate knowing Looks to propositional logic; impersonal way of knowing.	'I know these are equal, but maybe all base angles are not. I need a proof.'
Connected knowing Looks to what circumstances lead to perception; wants access to other people's knowledge.	'I know that it looks that way, but? What about the triangles that other people looked at? Let's look at those too.'
Constructed knowing Effort to integrate what is known intuitively and what other people know. Appreciates complexity of knowledge.	'Let's physically compare the angles. 'Tell me why you think that base angles are equal.'

Source: Belenky *et al.*, 1986

community colleges, and high-schools as well as in sites of informal education such as human-service agencies designed to support women in parenting. Based on hours of interviews with these informants, the authors propose 'stages' in knowing that differ in some fundamental ways from how men know. In doing so, the authors do not perceive these stages in the same way that we understand Piaget's stages. They are not necessarily meant as a developmental sequence through which all learners pass. These ways of knowing do, however, represent a progression from dependence to autonomy, from uncritical to critical. As with all theories, complete understanding of these stages will take time. In Table 20.1, I have attempted to illustrate the stages using the statement 'The base angles of an isosceles triangle are equal' and indicating what an individual might say in each stage. These statements give a glimpse of what is going on for the knower.

In the **Silence knowing** stage, knowing is subliminal; it does not belong to the individual and is usually not vocalized. All sources of knowing are external and come from authorities. The knower does not believe that she can learn from her own experience, and merely accepts or relies on an authority for all knowledge, which she does not question. Here, the inner voice of the knower would express an awareness that teachers think base angles are equal.

In the **Received knowing** stage, people learn by listening. Knowing comes from what authorities say, and the student depends on authorities to hand down the truth. Thus, knowledge is dependent upon an external source. There is no sense that the individual can create her own truths. Here, the knower would say, 'I know that base angles are equal because my teacher says so.' I frequently encounter received knowers among my college students. These are the individuals who, when asked why you cannot divide by zero, tell you the reason is 'my teacher told me'. They return the words of an authority. Learners in this category do not ask why a given rule is so, or wonder who gave their teacher the power to make such a decision. The authority that comes with being a teacher is all that is required for such students to accept the truth of a statement.

The **Subjective knowing** stage is a powerful one for the knower and legitimizes women's intuition. Here, knowledge derives from within, from that which feels right. Knowledge no longer comes from outside the knower, and an inner voice lets the individual know that she is on the right track. In this stage, the knower would say, 'I know that base angles are equal. Just look at them; they're equal.' Men and women handle this type of knowing differently. The male version asserts, 'I have a right to my opinion,' or 'It is obvious.' For women, there is a concern that their views do not intrude on those with opposing views. Their version is expressed as, 'It is just my opinion,' or 'I guess I feel so.'

For **Procedural knowing** to occur, a person usually requires some formal instruction or at least the presence of knowledgeable people who may serve as informal tutors and are able to model the process of providing evidence in support of an idea. At the procedural stage, the methodology used in providing evidence and presenting knowledge becomes important. In some instances, a particular methodology may be upheld as *the* way.

Two types of procedural knowing are identified in the book. **Separate knowing** is based on the use of impersonal procedures to establish truths. It is particularly suspicious of ideas that 'feel right' (as in subjective knowing). It often takes an adversarial form which is particularly difficult for girls and women, and separate knowers often employ rhetoric as if playing a game. The goal of separate knowing is to be absolutely certain of what is true. It is better to eliminate a possible truth than to accept as true something which later may prove to be false. The 'separate knower' would say, 'I know these are equal, but maybe all base angles are not. I need proof.' After this, she would go on to follow the rules of discourse in order to prove the statement.

Connected knowing, the second type of procedural knowing, builds on personal experiences. It explores what actions and thoughts lead to the perception that something is known. Experiences are a major vehicle through which knowing something takes place. Authority derives from shared experiences, not from power or status. A creative process would be used to gain experiences from which a

Table 20.2: Procedural Knowing

Separate Knowing	Connected Knowing
Logic	Intuition
Rigour	Creativity
Abstraction	Hypothesizing
Rationality	Conjecture
Axiomatics	Experience
Certainty	Relativism
Deduction	Induction
Completeness	Incompleteness
Absolute truth	Personal process tied to cultural environment
Power and control	Contextual
Algorithmic approach	
Structure and formality	

Source: Gilligan (1982)

conclusion could be drawn. The 'connected knower' would say, 'I know that it looks that way, but? What about the triangles that other people looked at? Let's look at those too.' In answering the question, 'Why do you think that?', the 'separate knower' would look to deductive logic. The 'connected knower', on the other hand, would want to know what circumstances led you to that conclusion.

It is in this stage, procedural knowing, that there is the most conflict with the traditional way of knowing in mathematics. If the only knowledge accepted as valid is that which can be statistically demonstrated or is based on deductive logic, methods that are independent of the knower's actions, then that which is known through induction would be devalued. And in mathematics, is not knowledge developed through deduction viewed as more valuable than knowledge gained through induction? But, how do we know what to set out to prove (separate knowing) if we do not first know things through inductive reasoning (connected knowing)? Let us look at some words we associate with these two kinds of knowing (Table 20.2). React to them as valid processes in knowing mathematics. If it is true that women tend to be 'connected knowers' and men 'separate knowers', and taking into account the way we value each of these ways of knowing and the way we teach mathematics, might this help account for the relatively small number of women pursuing mathematics-related fields? The authors have no 'hard data', to use a separate knowing term, to support the notion that women favour connected knowing, but such a preference would certainly fit into Gilligan's (1982) categorization of the conceptions of self. In fact, the authors of *Women's Ways of Knowing* have borrowed the words 'separate' and 'connected' from Gilligan.

The last stage in the model is **Constructed knowing**. Here, as the name implies, all knowledge is constructed by the knower. Answers are dependent on the context in which the questions are asked, and on the frame of reference of the asker. This type of knowing is particularly relevant to mathematics; for example, the solution of an equation is dependent on the domain being considered, and what can be proven is dependent on the axioms being assumed. It is in this stage that the learner can integrate her rational and emotive thoughts, as well as appreciate the complexity of knowledge formed from various perspectives. Here, the knower would use the creative aspect, induction, together with the rules of discourse, deduction, in order to know something. The constructed knower would say, 'Let's physically compare the angles. Tell me why you think that base angles . . .' There is a willingness to describe how the knowing occurred. In constructed knowing, the artificial dichotomy between separate and connected knowing, presented in Table 20.2, becomes apparent and the two ways of knowing can be merged, as recommended by Moody (1989) and others who think and write about feminist science. Certainly this category of knowing fits very well with current constructivist views on the teaching and learning of mathematics (Davis *et al.*, 1990).

Teaching and Learning Mathematics

Having outlined the stages proposed in *Women's Ways of Knowing*, the issue I would now like to explore is what the implications of this theory are for the teaching of mathematics. The authors make a plea for what they have termed 'connected teaching'. Connected teaching would address the issues important to the learning of women identified in the book, and thus might help students develop into constructed knowers. Some ideas for connected teaching are listed in Table 20.3 below, which I have adapted from Buerk (1985) and Morrow (1987).

In connected mathematics teaching, one would share the process of solving problems with students, not just the finished product or proof. Students need to see all the crumpled papers we put in the wastepaper basket, if they are to understand that mathematicians do not arrive at a solution the first time or the first way. This means that when we start teaching from a new mathematics book, we should not solve all the problems before class. Instead, we could show students how we start a given problem, make an error, and begin our solution over again. Female students need to watch women professors solve (and fail to solve) problems, and men professors fail to solve (and solve) problems. They need models of thinking that are human, imperfect, and most of all, attainable.

Mathematics needs to be taught as a process, not as a universal truth handed down by some disembodied, non-human force. Mathematics knowledge is not a predetermined entity. It is created anew for each of us, and all students should experience this act of creation. Presenting mathematics in the 'commercial with (male) voice-over' mode, as disembodied knowledge that cannot be questioned, works against connected knowing. The imitation model of teaching, in which the impeccable reasoning of the professor as to 'how a proof should be done' is presented to students for them to mimic, is not a particularly effective means of learning for women.

In connected teaching, teacher and students would engage in the process of thinking and discovering mathematics together. Alternate methods of solution

Table 20.3: Examples of connected teaching in mathematics

Issue	Importance for connected teaching	Connecting teaching in mathematics
Voice	• Education occurring in the context of conversation • Gaining a sense of self • Developing one's own authority	• Students develop own justification • Group problem-solving • Writing exercises • Confidence-building workshops • Class discussions • Student-designed projects • Teacher does not give answers
First-hand experience	• Building on intuitive understanding • Validating students' knowledge base • Providing insight into reasons for area of study • Encouraging activity versus passivity	• Applications-oriented workshops • Drawing and constructing models • Using visual representations • Using computers
Confirmation of self as knower	• Providing a basis for area of study • Becoming a rule maker • Becoming a constructor of knowledge • Providing a basis for midwife model of education	• Looking for what is already known • Listening to students' reasons • Respecting students' ideas • Engaging class in outside activities
Problem-posing	• Focusing on process rather than outcomes • Allowing students to see uncertainties and understand knowledge construction	• Allowing students to struggle for solution • Having high expectations • Focusing on explanation • Sharing our problem-solving
Believing versus doubting	• Giving students alternative modes for discourse	• Asking for right explanations even when right • Answering questions with questions
Support versus challenge	• Allowing students to become independent learners	• Validating present understanding while providing challenges
Structure versus freedom	• Giving guidance/mentoring without imposing tyrannical expectations	• Following prepared curriculum while allowing for explorations

Source: Adapted from Buerk (1985) and Morrow (1987)

would be encouraged. In problem-solving, for example, more emphasis would be placed on finding another way to solve a particular problem, than on solving additional problems in the same manner. Students would use their intuition in an inductive process of discovery. Here, the emphasis would be on generating hypotheses. For example, many polygons could be examined to determine the sum of their interior angles and in this way generate a formula for the sum of the interior angles of any polygon. Patterns could be used to discover why the product of two negative integers is a positive integer. The generation of these relationships would be valued as much as their verification, for without this prior inductive work, we would be at a loss to know what to prove.

A metaphor used by the book to describe connected teaching sees the teacher as a midwife, nurturing the student's newborn thought in a supportive environment. The first concern of this midwife/teacher is to preserve the student's fragile, newborn thought. The midwife/teacher then supports the evolution or growth of this thought. The goal is not to replace it with a different, teacher-generated, 'better' thought. Rather, it is to help the student's thought grow, mature, and develop. Both the teacher and the student engage in this process. Their roles merge, and the teacher of students and the students of the teacher cease to exist while a new team emerges: teacher–student and students–teachers (Freire, 1971).

In connected teaching, groups are created in which members can nurture each other's thoughts to maturity. Here, no one needs to apologize for uncertainty. Questions are posed and potential answers are explored together. More courses would be conducted in a style of community, and fewer would be taught in the masculine style of adversarial discourse or hierarchical pontification. Truth or knowledge is constructed through consensus, not through conflict. Diversity of approach is welcomed in discussions. Connected teachers trust their students' thinking and encourage them to expand upon it. This is particularly important because the women reported in the Belenky *et al.* (1986) study experienced doubt as debilitating rather than energizing.

Recent research provides evidence that females prefer and do better in mathematics classes which use cooperative learning in small groups, rather than individualized or competitive learning strategies (Fennema and Leder, 1990). Moreover, as discussed above, a collaborative approach seems to offer more possibility for connected teaching. Certainly, a small-group environment provides an atmosphere in which different points of view can be shared and students can discuss mathematics. And it is difficult to envision how students can construct their own knowledge by weaving together objective and subjective knowing in a mathematics class taught in a traditional teacher-centred lecture format.

There is considerable evidence that mathematics teaching at all levels neither exemplifies connected teaching, nor encourages constructed knowing (National Research Council, 1989; Weiss, 1989). How much difference might such a change in the teaching of mathematics make in the improving attitudes and performance in the subject for more students of both sexes? Some beginning work in this direction is positive (Buerk, 1985; Damarin, 1990; Rogers, Chapter 21, this volume). But, I suggest that much more research is needed to help us identify how best to encourage connected teaching, and to determine what impact it might have, especially on women's participation in mathematics-based careers.

Questions for Future Research

In this section, I will suggest some avenues for future investigation that revolve around *Women's Ways of Knowing*. This is a fertile area to study for it might take us beyond the research of the last fifteen years on gender differences in mathematics education by providing a different theoretical model with more explanatory power.

First, I have reanalysed data collected before the appearance of *Women's Ways of Knowing* (Becker, 1990, 1984). In this research, I conducted in-depth interviews with male and female graduate students in the mathematical sciences in order to identify factors associated with their decision to attain a graduate degree. Thus, I was focusing on the point between undergraduate and graduate study in mathematics at which we lose many women (see Friedman, Chapter 5, this volume). There were no questions in my protocol which specifically addressed aspects of the *Women's Ways of Knowing* model. However, one particular portion of the interviews focused on the period at which students' interests in mathematics developed, as well as on what students liked about the subject. Although there were no gender differences apparent in these categories, several factors which relate to *Women's Ways of Knowing* are worth mentioning here. Beyond liking mathematics because they were good at it, graduate students were attracted by the problem-solving aspect of the subject, and particularly liked starting with certain assumptions and solving puzzles logically. The objective nature of the discipline, as described by the informants, also appealed to them. They liked being able to tell if a problem was solved or if a proof was correct. This observation leads to at least one area for further research. Are women in mathematics more likely than non-mathematicians to be separate knowers and thus be attracted to the subject because, at least at this early stage, they perceive mathematics to be an objective discipline in which they can find absolute truth? Do their views of mathematics evolve as they pursue further study and actually do research themselves?

A second factor I observed in my studies was that nearly all informants developed their liking of mathematics quite early, and before high-school. Frequently, a teacher was mentioned as the one who piqued their interest by providing an enriched curriculum which went beyond arithmetic to include problem-solving or topics in algebra. Thus, there does seem to be the possibility that teachers and methods of instruction might make a difference in students' ultimate career choices. Could an extensive use of connected teaching affect more students in this positive way?

Other questions which arise from the model as described in this chapter include the following:

- Does the model represent how women (and men) come to know mathematics? How might we proceed in testing this model in mathematics?
- Do more men fall into the separate knowing category and more women into the connected knowing category? Is there any relationship between

these categories and career choices? How might one begin to ascertain the type of procedural knowing category into which an individual fits?
- Can the model help to explain the continued underrepresentation of women in mathematics-based fields?

Because I believe that *Women's Ways of Knowing* has important implications for the teaching of mathematics, additional questions relating to this area arise. This book does provide evidence that the traditional style of teaching mathematics is in conflict with how many women learn. Connected teaching, which emphasizes connection over separation, understanding and acceptance over assessment and doubting, and collaboration over debate, may be a more effective technique. We need to investigate questions such as these:

- Can a change to connected teaching at all levels reclaim mathematics and science for women students who have already been turned off the subjects? For example, instead of giving up on women already in college who are choosing fields which do not require mathematics, can we attract them to mathematics and science by good teaching in entry level courses? Such a strategy has been suggested by Tobias (1990).
- Does connected teaching make a difference in student performance and in attitudes towards mathematics?
- How can we help mathematics teachers at all levels become constructed knowers and connected teachers?

Conclusion

As mathematics educators continue to investigate ways to increase women's participation in mathematics and mathematics-related careers, the ideas in *Women's Ways of Knowing* should be given considerable weight. In so doing, the mathematics education of all students might be enhanced by the wider application of connected teaching. If the way mathematics is currently taught alienates many women because it does not appreciate or validate their ways of knowing, then many women may choose not to pursue mathematics and mathematics-related careers.

What are the political implications of broaching the possibility of a different way of knowing? Should we fear that some will use this difference as a means of proving women's inferiority? There is always a danger that knowledge can be used against people, but there is a greater danger in not addressing this issue at all. If there is the chance that connected teaching could enhance the education of women, then it is an approach that should be seriously considered. If movement from subjective to connected knowing is based on experience, and if connected knowing is women's preference, then, and assuming we want them to know, we must present information to women in a different way and accept another way of knowing. To ignore these differences in how women and men come to know is to deny large numbers of people access to mathematical knowledge. Addressing the implications of *Women's Ways of Knowing* will not only benefit female students, but society at large. Who can say what that roar which lies on the other side of silence might accomplish in mathematics?

Note

1 Many of the ideas in this chapter have been developed through discussion with Judith E. Jacobs (see, for example, Jacobs (1994)). The author acknowledges her inspiration to pursue this topic, her scholarly thoughtfulness, and her collegial sharing of ideas. Portions of this chapter have been published by Jacobs and Becker in the newsletter of Women and Mathematics Education and an earlier version was discussed at a Symposium of IOWME in June 1991.

References

BECKER, J.R. (1984) 'The pursuit of graduate education in mathematics: Factors that influence women and men', *Journal of Educational Equity and Leadership*, 4, 1, pp. 39–53.

BECKER, J.R. (1990) 'Graduate education in the mathematical sciences: Factors influencing women and men', in BURTON, L. (Ed) *Gender and Mathematics: An International Perspective*, London, Cassell, pp. 119–30.

BELENKY, M.F., CLINCHY, B.M., GOLDBERGER, N.R. and TARULE, J.M. (1986) *Women's Ways of Knowing: The Development of Self, Voice, and Mind*, New York, Basic Books Inc.

BUERK, D. (1985) 'The voices of women making meaning in mathematics', *Journal of Education*, 167, 3, pp. 59–70.

DAMARIN, S. (1990) 'Teaching mathematics: A feminist perspective', in COONEY, T.J. and HIRSCH, C.R. (Eds) *Teaching and Learning Mathematics in the 1990s: 1990 Yearbook*, Reston, VA, NCTM, pp. 144–51.

DAVIS, R.B., MAHER, C.A. and NODDINGS, N. (1990) *Constructivist Views on the Teaching and Learning of Mathematics*, Reston, VA, NCTM.

ELIOT, G. (1871–2) *Middlemarch*, New York, Penguin Books.

FENNEMA, E. and LEDER, G. (Eds) (1990) *Mathematics and Gender*, New York, Teachers College Press.

FREIRE, P. (1971) *Pedagogy of the Oppressed*, New York, Seaview.

GILLIGAN, C. (1982) *In a Different Voice: Psychology Theory and Women's Development*, Cambridge MA, Harvard University Press.

JACOBS, J.E. (1994) 'Feminist pedagogy and mathematics', *Zentralblatt für Didaktik der Mathematik*, 26, 1, pp. 12–17.

MOODY, J.B. (1989) 'Women and science: Their critical move together into the 21st century', *NWSAction*, 2, 2, pp. 7–10.

MORROW, C. (November, 1987) 'Connected Teaching in Mathematics', Paper presented at the annual meeting of Research on Women and Education, Portland, OR.

NATIONAL RESEARCH COUNCIL (1989) *Everybody Counts: A Report of the Future of Mathematics Education*, Washington, DC, National Academy Press.

PERRY, W. (1970) *Forms of Intellectual Development in the College Years*, New York, Holt, Rinehart, and Winston.

TOBIAS, S. (1990) *They're Not Dumb, They're Different: Stalking the Second Tier*, Tucson, AZ, The Research Corporation.

WEISS, I. (1989) *Science and Mathematics Briefing Book*, Chapel Hill, NC, Horizon Research.

Chapter 21

Putting Theory into Practice

Pat Rogers

Listen to a woman groping for language in which to express what is on her mind, sensing that the terms of academic discourse are not her language, trying to cut down her thought to the dimensions of a discourse not intended for her. (Rich, 1979)

Introduction

A forceful impetus for changing my teaching has been reflecting on my own experiences in learning undergraduate mathematics and doing mathematical research.[1] When I first started teaching, I adopted those practices I had observed as a student: I lectured. I believed that teaching at the post-secondary level involved the transmission of knowledge from me, the *expert*, to the students, the *novices*. I saw my job as exposing the students to the content of the course. The students' job was to master this material by listening attentively as I explained ideas, by watching carefully as I showed them how to solve problems, and by practising solving problems on their own at home. I administered tests in order to measure achievement and to rank students in relation to their peers. Each lecture had a natural pattern: I introduced a topic, covered the blackboard with formulas and mathematical language, and worked a couple of illustrative examples. I asked a few questions. I even elicited a few answers — usually from the same three or four (male) students. Finally, I assigned homework. I was considered a successful teacher. My course evaluations proved it. Students praised me for my enthusiasm, my organization, the clarity of my exposition, my knowledge of the material and my accessibility. The most critical comment was a request to slow down a little. Yet, when the final examination came, many students failed or wrote such incomprehensible answers that I wondered if we had all been involved in the same course. How could they do so badly, I wondered, when I had explained the material so well? How could they not buy the goods I had sold them so persuasively? Of course this troubled me, but it was easy enough to dismiss: 'Students come to university so ill-prepared. If only we had better students, just think what we could do!' A familiar refrain?

Over the years I became increasingly concerned about the students in my class who were silent. These were students who sorely needed individual attention yet never used my office hours. These were the students who were heading for certain failure but were lulled into thinking they might pass the course because, as they often said, 'It seems so easy when you do it on the board.' The majority of these students were female. I was able to experience their private distress only when I began to reflect on my own experiences as a mathematics undergraduate at Oxford. It shocked me to realize how faithfully I was reproducing in my own classroom the structures which had so effectively silenced and disempowered me then.

It seems to me there are two important questions to consider when examining the ways in which traditional mathematics pedagogy at the post-secondary level may disempower students. Whose mathematics are we teaching? How are we teaching it?

Whose Mathematics Are We Teaching?

As a high-school student, I was able to reconstruct for myself any formulas, rules or results that I needed. Some of my teachers even admitted to not knowing answers to many of the problems that I posed. However, at the university the professor was distant, remote and expert and I began to experience mathematics as something distant and imposed, not something that *I* had a role in producing. We were rarely given the opportunity to play with mathematical ideas or to construct our own meanings (except on our own at home). Instead, through the medium of the polished lecture (or textbook), mathematics came to me finished, absolute and pre-digested.

A pedagogy that emphasizes 'product' deprives students of the experience of the 'process' by which ideas in mathematics come to be. It perpetuates a view of mathematics in which right answers are the exclusive and sole property of experts. Such a pedagogy strips mathematics of the context in which it was created and reinforces misconceptions about its very nature. Students permitted to see only the polished product may come to believe they can never create similar results for themselves ('I didn't know where to begin!').

Carol Gilligan's (1982) research in the field of women's moral development provides a link between 'math avoidance' and mathematics pedagogy.[2] She identified two styles of reasoning, which although not gender-specific are thought to be gender-related: one, the traditional style, is characterized by objectivity, reason, logic, and an appeal to justice; and the other, 'the different voice', often identified with women and as a consequence devalued, is characterized by subjectivity, intuition, and a desire to maintain relationships. The 'different voice' is rooted in an ethics of care and responsibility.

Dorothy Buerk (1985) has studied Gilligan's styles in relation to the work of mathematicians. She presented a group of mathematicians with two lists of words describing examples of 'separate' reasoning (the traditional style) and 'connected' reasoning (the different voice) in the field of mathematics. They unanimously identified the 'connected' list as representing the way mathematicians do mathematics. '"[M]athematics is intuitive", they said. They stressed the creative side: attention to the limitations and exceptions to theories, the connections between ideas, and the search for differences among theories and patterns that appear similar. And yet they agreed that the "separate" list conveyed the way that mathematics is communicated in the classroom, in textbooks, and in their professional writing' (Buerk, 1985).

In reality, mathematicians employ *both* forms of reasoning in their work. But the problem with mathematics teaching, particularly at the post-secondary level where the lecture mode of instruction is so predominant, is that the creative-intuitive form is largely eliminated. Students are not given the opportunity to be involved in the process of constructing mathematical ideas, a process in which 'connected' thought is so important. There is an enormous cognitive gulf between the way mathematics is presented and the individual ways in which it is possible and natural to arrive at an understanding of it, or to construct it for oneself. Some students are able to bridge this gap for themselves, but many are not.

Of course, the expository approach to teaching mathematics affects *all* students, but it does not affect them all to the same extent. If Gilligan's claim that women favour a 'connected' reasoning style has validity, then the tendency of most teachers to take only the traditional 'separate' route to mathematics could inhibit or prevent some students — especially women students — from claiming mathematics for themselves.

How Are We Teaching Mathematics?

One important component of the expository method of teaching is the use of the distant authority to 'impart knowledge'. Practices, such as lecturing, subordinates students' knowledge and understanding to that of the professor and the even more distant authority, the textbook. Solving problems *for* students does not teach them to solve problems for themselves. Instead, it disempowers students by rendering them passive, and conveys mistaken notions such as, 'there is only one correct solution, namely the teacher's'. It can also give students the impression that the teachers lack confidence in their ability to solve problems for themselves, that is, to think mathematically. Such implicit judgments of students' abilities as well as the more explicit verbal and written judgments they receive can stifle students' emerging styles and diminish their confidence to engage. Indeed, while students do need to acknowledge that they will make mistakes, they also need to know that they are intrinsically capable. According to the findings of Belenky *et al.* (1986), this may be particularly important for women students. The unconscious or uncritical use of authority in the classroom is a powerful disabler.

The fact that women do tend to be more silent in the classroom is well documented. Belenky *et al.* (1986) point to the difficulty with the development of 'voice' in situations dominated by male authorities. Krupnick (1985), in her study of gender in the classroom at Harvard College, found that women speak less often than male students, tend not to compete with them, and are more sensitive to interruption (they do not pursue their thought once they are interrupted and they are interrupted more often than male students). Inequitable access to classroom discourse makes it very difficult for women to cultivate and make use of an original voice. If alienation from the subject results when a student's 'sense' or experience remains unexpressed and/or uncultivated, if alienation can be thought of as a muting of the self (Lewis and Simon, 1986), and if the ability to articulate an idea signals ownership of that idea, then silence in the mathematics classroom spells a serious limitation in the ability to claim a subject freely and with energy and self-confidence. I suggest that without this sense of ownership, the optimal development of mathematical thinking skills may not take place. Creative, subjective engagement will be curtailed, and therefore those thinking skills dependent on such engagement may remain underdeveloped.

A Mathematics Pedagogy of Possibility

I would like to outline an alternative approach to the teaching of mathematics, one which is rooted not in the authoritative and imposed style which distances and silences, but in a style which encourages direct access and engagement, free creative expression, and ownership of the subject. This alternative might be thought of as introducing a feminist element.

The first element of this approach has been influenced, among other things, by an ethics of caring and responsibility that is designed to enable those who have been silenced to speak (Simon, 1987). One must become, first and foremost, a 'caring teacher'. Here I use the term caring not simply as a feeling of concern, attentiveness or solicitude for another person, but in the very specific sense of 'caring' as in helping 'the other grow and actualize himself' (Mayeroff, 1971). In this view, caring teachers are able to focus primarily, or at least initially, upon the student rather than upon the subject matter, with the idea that the route to the subject is not imposed from without, but rather illuminated from within. They do not work upon their students, but with their students, looking at the subject matter from their perspective and at their level. Elsewhere, I have called this a student-sensitive pedagogy (Rogers, 1988), for it is grounded in the students' own language, is focused on process rather than on content, and centres on the students' individual questions and learning processes. Students who are 'cared for' in this way are set free to pursue their own legitimate projects (Noddings, 1984).

Another element in this approach involves demystifying the doing of mathematics. This includes such things as calling students' attention to mathematics as a creation of the human mind, making visible the means by which mathematical ideas come into being and the process by which they are 'polished' for public consumption. It also includes encouraging students to challenge authoritative discourse, and to feel that the gates to the mathematical community are open and that they have the skills they need to walk through and to operate within it.

Reflecting on the differences between what I was doing in my early teaching and what I came to believe needed to be done, I realized that a core difference lay in the engagement of students *within* the classroom in purposeful, meaningful activity.

A Feminist Mathematics Pedagogy in Practice

In this section, I describe some of the ways I have put the ideas outlined above into practice. At the beginning of a course I discuss my goals with the students and make the goals available in written form. They fall into three categories: content-specific goals, process goals, and communication/social goals. The content-specific goals include the development of students' skills in constructing mathematics and writing their own proofs clearly and precisely in convincing form. Process goals include the development of independent working skills (the more highly developed the individual working skills, the less the need for a teacher). Some of these skills include reading a mathematics text with understanding; finding, analysing and correcting mistakes; and asking mathematical questions. Since mathematics is an activity which depends on communication between colleagues, social goals are also stressed and include helping students collaborate with others; communicating ideas clearly and with confidence both orally and in writing; listening actively and offering constructive criticism; asking and responding to challenging questions. My teaching methods vary depending on the characteristics of the particular class and the questions raised by the students. Following is a brief description of some of the methods I use most frequently.

Lecturing

I have not completely abandoned lectures; instead I use them sparingly for three main purposes: to introduce a new section of material; to conclude a topic and draw everything together; to introduce a new concept and to motivate assigned readings and problems (this is usually a short lecture given at the end of a session).

Think–write–pair–discuss

The 'think–write–pair–discuss' idea is adapted from Davidson *et al.* (1986) and is the most useful and most frequently used of my current strategies. It is also the one that is most adaptable to student groups of all sizes. I present students with a question, or give them a segment of the text to read. They work independently at first, put their thoughts and ideas down in writing, and then form pairs to discuss. This provides support for those students who are unsure of their ideas or who have a fear that what they say might appear foolish in public. It has the effect of increasing and improving participation and involving all students. This may precede or be integrated into all the activities which follow.

Whole-group dialogue
Like a lecture, whole-group dialogue is teacher-directed. Unlike a lecture, it is student-centred. Dialogue usually follows assigned reading and think–write–pair–discuss activity, and operates by way of questioning. My questions aim to help students express their ideas clearly and with precision. For this, good active listening skills on my part are essential: I demand reasons for statements, and challenge students to formulate their ideas in their own words and explain them to each other. The latter is important because students are often unable to 'hear' their instructor, but are perfectly able to hear one another, and such listening and telling is empowering.

Board work
Students come to the board individually or in small groups to write up questions, problems, and solutions. We then discuss the solutions. I may assign problems ahead of time, and I may allow time in class for students to work on solutions in groups prior to the board work. Sometimes, several groups will work on the same problem at the board. I am then able to demonstrate that there is no single right approach to solving a problem. Students learn from each other and learn both oral and written presentation skills. At first, I do not require that students remain at the board to explain their solution and answer questions. However, as the classroom climate becomes more supportive and students begin to encourage each other to participate, confidence in their ability to discuss mathematical ideas increases.

Brainstorming
I give student pairs between two and five minutes to write down everything they know about a given topic. I then call on the pairs and all the information is put on the blackboard. Another way of gathering the information is to ask questions of each student in turn, giving students having the right to 'pass' — I call this technique 'round robin'. Discussion then centres around classifying and evaluating what has been gathered. I use this technique most often for review or to begin an investigation (see below).

Problem-posing
Students generate their own questions in a particular area. These are then examined, and the students commit themselves to a particular conjecture. For example, one year an early conjecture in my group theory course was Lagrange's theorem. Proving this theorem became the focus of our work for several weeks. This focus made the course problem-driven, a response to the students' own agenda and chosen avenues of inquiry. An activity of this sort usually occurs naturally, unlike an investigation which is artificially established.

Investigation
The class examines patterns found in concrete examples in order to uncover algebraic structure. Generalizations are then made in the form of conjectures, and a theory which proves or disproves the conjectures is developed. I have used investigations to generate all standard structure theorems for finite groups.

Small-group work

Groups of two to four students work together while I circulate among them and check their work. This provides me with immediate feedback on student understanding and enables students to influence the pace and the development of the course. Even with large groups of over 100 students, where I can check only the work of the students at the aisle, I am provided with immediate feedback.

Reading exercises

The development of good reading skills is essential to gaining independence in mathematics. An effective way to help students develop good reading skills is to insert question marks at crucial junctures in a proof before assigning it to students for reading. I ask students to rewrite the proof, replacing each question mark with an explanation. I vary this activity after the students have gained some skill by having them insert their own question marks. To give students more practice in 'talking mathematics' I may ask them to give the explanations for the question marks orally either to the whole group, or by having them pair up as in 'think–write–pair–discuss' — this is a particularly useful way to help students develop skill in understanding definitions. When students have more confidence I have them pair up to pull apart the statement of a theorem and write it in their own words prior to proving the result. All technical words must be defined. The structure of the statement is examined and a proof strategy outlined. This is often done in conjunction with the 'forward–backward proof' technique (see below).

Proof generation

The 'forward–backward proof' technique described by Solow (1982) in his book, *How to Read and Do Proofs*, is a very empowering strategy. It takes its name from the way mathematicians typically organize their thoughts when attempting to prove a conjecture. Although it cannot be successfully applied in all situations, it has virtually eliminated complaints from students in my courses that they 'don't know how to begin!'

Community

A classroom climate characterized by safety and trust is essential for risk-taking to occur and for students to be willing to test their ideas and thoughts and develop fluency in mathematical language. In other words, I work to build a community of mathematicians.

Some students have difficulty adjusting to my approach to teaching because it includes them constantly and individually. For this reason, constant monitoring of my teaching is important. I do this primarily through the use of 'one-minute papers' (Wilson, 1986). The one-minute paper is a simple device for obtaining immediate feedback on the effectiveness of teaching. In one version, I ask students to take out a piece of paper in the last minute or two of a session and write down one or two points they are still confused about. They hand these in anonymously and I use their responses to plan the next session. Alternatively, I may ask the students how they feel about various aspects of my teaching. For example, through this device I learned that one student found my questioning techniques and practice of asking her to come to the board were intimidating and humiliating for her. She did not like being singled out, fearing the public display of her perceived ignorance. On the other hand, she admitted that she appreciated the opportunity of learning from other students. Through negotiation, we were able to arrive at a compromise which made her, and many other students in the class, more comfortable about participating. The result was that I began to use the 'round-robin' technique (in which students are allowed to 'pass') more frequently and have students come to the board in pairs. In mathematics, once a climate of trust has been firmly established, these one-minute papers can be used very informally during the session. For example, I might ask students to work through an exercise which will demonstrate whether they have understood a concept we have just explored. While they are writing, I walk around and check their work to see how well they have understood the topic before proceeding to a new one.

Evaluation of Student Learning

I evaluate students' learning constantly in the sense that the student-centred teaching techniques and the one-minute papers I employ provide immediate, frequent and regular feedback on students' understanding of the course content and processes. Following is a brief description of some of the more formal evaluation methods I use.

I assign *homework* regularly throughout the course and space these assignments so that, when combined with quizzes and examinations, students receive regular feedback on three kinds of written work. I encourage students to collaborate on assignments but accept only independent write-ups. I comment on their work but do not grade it because I want them to experiment freely with the concepts and processes of the course without being penalized for doing so.

Students can earn *participation credit* in a variety of ways, allowing for their individual learning preferences. Some of these ways include visiting me for an office consultation, presenting aspects of assigned readings in class, participating in assigned collaborative exercises, sharing ideas by coming to the blackboard and presenting a proof, asking questions, offering explanations, or joining in discussions. The issue of participation credit in the final-grade calculation is a matter of some concern to me. I have used it often in the past but have found that it can foster competition, regardless of the quality of classroom climate I have been able to establish.

I give *tests* regularly, but only in conjunction with giving students opportunities for improving their grades but more importantly their comprehension of the course material. When I first assess a student's work, I simply indicate where the argument breaks down and award a mark, with no explanation. Occasionally, I ask a question calling for clarification. The student then responds to me in writing. The importance of giving students a second chance far outweighs its value in increasing their grade. An early failure in a course can be very dispiriting. As well, I believe that tests should and can be a vehicle for learning, and not solely an instrument of evaluation. Many students tell me that they appreciate the opportunity to learn from analysing their own work. I have also found that it improves the learning environment in the class and reduces students' test anxiety.

I also believe in formal *examinations*. Formal examinations are like final performances, an opportunity for students to pull the course together, to see it as a whole and to demonstrate how much they have learned. I have experimented, using cooperative learning techniques (see, for example, Aronson *et al.*, 1978), with giving students responsibility for generating final examination questions. The result of this experiment in collaborative examination design was that it fostered a cooperative spirit among the students towards preparing for the examination — indeed they probably learned more from this activity than from the examination itself.

Conclusion

My first experience in employing participatory teaching methods dates back to a second-year finite mathematics course. Two other parallel courses were taught that year by other instructors using traditional lecturing methods. My main objective then was to prove to my colleagues that I could teach sixty students without lecturing, without compromising standards, and without disadvantaging students either because I 'covered' less material or because they achieved lower final grades than students in the two parallel courses (these are the criticisms commonly made by mathematicians in regard to those who abandon lecturing). The result was encouraging and convincing.

Measured by a common final examination, the three courses had almost identical averages. However students in my course obtained more of the highest grades overall and far fewer of them failed. Covering the course content presented no difficulty, either for me or for the students, judging by their examination success. Over two-thirds of the students participated actively in the course in one way or another and only one student complained of any discomfort with my teaching methods.

Judged by examination criteria alone, one cannot conclude that the lecture method came out second best. However, it is clear that in this first try, my approach held up well against the traditional approach. On the other hand, it is questionable whether the impact of participatory, democratic teaching methods can readily be evaluated in the short-term, or by quantitative means. I was in fact more interested in discovering whether I *could* teach this way and whether students would be responsive to these methods and learn from them. This took precedence in my mind over comparing these techniques with conventional methods or trying to prove their superiority.

I return now to the project with which I began this discussion, that of developing a pedagogy designed among other things to suit women's styles of learning. Since that first experiment I have further refined my approach to participatory teaching in both advanced mathematics courses and in courses for pre-service teacher candidates. Whatever the level of the students, whatever their background preparation in mathematics, however advanced or elementary the mathematics included in the course, the results have been the same. My courses enjoy high demand, lower than usual levels of attrition and absenteeism, and higher and more consistent achievement. What is more important to me, however, is the satisfaction students exhibit as they become fluent and competent with the process of doing mathematics. A consequence of teaching in this way, a consequence I have found more rewarding and compelling than any other, is the strong sense of community and caring that develops among students. Students are more concerned about each other's welfare and progress than anxious to compete with each other for higher grades. Students develop close bonds of loyalty to one another which carry well beyond the course itself and are often long-lasting. Lack of competition is not usually associated with the mathematics classroom (however, see Rogers (1988)), yet it appears to be a very natural outcome of my teaching now, and evidently provides a more comfortable and equitable classroom environment for the female students I teach.

Now that I know I *can* teach this way, I discover that I prefer teaching this way. It allows me to know the students well and I am now much more aware how they comprehend the material — I do not have to wait until the final examination to find that out, as is so often the case when one lectures. By providing opportunities for students to hear and to develop their own voices through engagement in authentic mathematical activity within the classroom, I am able to engage them in purposeful, meaningful academic discourse, allowing them to claim ownership of mathematics for themselves. In so doing, I believe I not only avoid discriminating against students who are currently denied access to mathematics (especially women), but I also provide a more meaningful and equal mathematics education for *all* students.

Notes

1. This chapter is a revised version of Rogers (1992).
2. I use the term 'math avoidance' because it is well known. However, my own view is that society has chosen to exclude women from the study of mathematics. Labelling women's absence from mathematics 'avoidance' is a monocultural view because it fails to recognize that it is the discipline (in both senses) of mathematics itself which turns women away. Women may indeed choose to avoid mathematics, but the term 'math avoidance' does not connote a deliberate choice.

References

Aronson, E., Blaney, N., Stephan, C., Sikes, J. and Snapp, M. (1978) *The Jigsaw Classroom*, Beverley Hills, CA, Sage.

Belenky, M.F., Clinchy, B.M., Goldberger, N.R. and Tarule, J. M. (1986) *Women's Ways of Knowing: The Development of Self, Voice and Mind*, New York, Basic Books.

Buerk, D. (1985) 'The voices of women making meaning in mathematics', *Journal of Education*, 167, 3, pp. 59–70.

Davidson, N., Agreen, L. and Davis, C. (1986) 'Small group learning in junior high school mathematics', *School Science and Mathematics*, pp. 23–30.

Gilligan, C. (1982) *In A Different Voice: Psychological Theory and Women's Development*, Cambridge, MA, Harvard University Press.

Krupnick, C.G. (1985) 'Women and men in the classroom: Inequality and its remedies', *On Teaching and Learning*, 1, pp. 18–25.

Lewis, M. and Simon, R. (1986) 'A discourse not intended for her: Learning and teaching within patriarchy', *Harvard Educational Review*, 56, 4, pp. 457–71.

Marcroft, M. (1990) 'The politics of the classroom: Toward an oppositional pedagogy', *New Directions for Teaching and Learning*, 44, Winter, pp. 61–71.

Mayeroff, M. (1971) *On Caring*, New York, Harper and Row.

Noddings, N. (1984) *Caring*, Berkeley, CA, University of California Press.

Rich, A. (1979) *On Lies, Secrets, and Silence: Selected Prose, 1966–1978*, New York, Norton.

Rogers, P. (1988) 'Student-sensitive teaching at the tertiary level: A case study', in *Proceedings of the Twelfth Annual Conference of the International Group for the Psychology of Mathematics Education*, 2, pp. 536–43.

Rogers, P. (1990) 'Thoughts on power and pedagogy', in Burton, L. (Ed) *Gender and Mathematics: An International Perspective*, London, Cassell, pp. 38–46.

Rogers, P. (1992) 'Transforming mathematics pedagogy', *On Teaching and Learning*, 4, pp. 78–98.

Simon, R. (1987) 'Empowerment as a pedagogy of possibility', *Language Arts*, 64, 4, pp. 370–82.

Solow, D. (1982) *How to Read and Do Proofs: An Introduction to Mathematics Thought Processes*, New York, J. Wiley.

Wilson, R.C. (1986, March/April) 'Improving faculty teaching: Effective use of student evaluations and consultants', *Journal of Higher Education*, 57, 2, pp. 196–211.

Chapter 22

Gender Reform through School Mathematics

Sue Willis

Introduction

Feminists, from a variety of perspectives, have argued that aspects of school mathematics disadvantage girls. Some adjustments to curriculum have occurred in order to make school mathematics less overtly sexist and more attractive to girls, and some exciting and innovative curriculum projects have been developed. Nevertheless, certainly in Australia, and I believe elsewhere, these developments tend to remain marginal to the mainstream curriculum, while major mathematics-curriculum reforms occur with little serious attention to matters of gender. While many feminists continue to critique school mathematics, on the whole the mainstream mathematics curriculum remains unchallenged and relatively unscathed by feminism. Even those of us who are most concerned with girls and mathematics have tended to focus on access as the problem, the task being 'to get it rather than change it' (Johnson, 1983, p. 15).

Changing Girls, Changing Choices

Reform programmes that target girls and mathematics often have focused on changing girls by improving their achievements, attitudes, confidence, self-esteem, and risk-taking behaviour, or on altering their educational or career choices by improving their participation in mathematics or mathematically related occupations (Kenway and Willis, 1993). The former include programmes intended to enhance self-esteem with respect to mathematics, coupled with courses aimed at building confidence and competence in mathematics. The latter include the use of role models, workshops designed to 'motivate' mathematics, educational and career support and advice, all intended to encourage girls to make pro-mathematics choices. As important as many of these programmes have been and may continue to be, in my view reforms that have as their main focus changing girls or changing their choices will have limited success in enhancing female life options. Let me sketch out some of my reasons for this view.[1]

First, they suggest a deficit view of girls — that there is something lacking or in need of reconstruction — even when the intention is to value and affirm girls and girlhood. In order to argue disadvantage and justify intervention, we often find ourselves emphasizing difference in a form that negatively compares girls to boys, women to men. In mathematics, girls *lack* self-esteem or confidence, they *fear* success, they make *poor* choices, they *do not* understand or act in their own best interests, they achieve *less* well — all relative to boys. Thus, we perpetuate certain ways of viewing females and males which undermine girls and women. Indeed, girls have the right to question the extent to which many of us, who are supposed to have their best interests at heart, remain silent in the presence of such subtle undermining. Male skills and attributions are seen as the norm against which females become the negative other — to become equal is to become male.

Second, implicit in many of these approaches is the view that the relative educational disadvantage of girls causes their occupational disadvantage. That is, gender inequalities in the working world can be overcome if sufficiently many girls leave school with the necessary mathematical skills and attributions to take their full share of the 'good jobs', or if they change their individual aspirations and decisions. That girls should plan for economic independence and consider the full range of choices which may be available is clear. That their choices about mathematics should not be constrained by their gender is also. However, in attempting to encourage more girls and young women to participate in mathematics, we often paint a rather rosy picture of the advantages which will accrue to those who do so. For it is not at all clear that, if girls and young women made the same mathematics and career choices as boys and young men, their earnings, status, and level of employment would be the same. Horizontal desegregation of the labour market may be desirable, but as long as vertical segregation remains in place, women are likely to be worse off in male-dominated fields (Kenway, Willis and Junor, 1994).

My third concern is closely related. Encouraging in girls a sense of agency is essential, and it is here that the real strength of many programmes which endeavour to change the ways girls think about their lives is found. To suggest or imply that girls have a 'free (unconstrained) choice' to be and do anything, and that their 'disadvantage' can be overcome by changes in their individual attributions or choices about mathematics is, I believe, dishonest and possibly even cruel. The options of girls struggling with poverty, racism, isolation or disability are often quite limited, regardless of individual qualities or decisions. Furthermore, all of our choices are constrained by cultural norms and values associated with gender, race, and class. Around the world, women are still regarded as the primary care-givers, and girls and women shoulder the major responsibility for tending the social fabric. In Australia, the struggles of women and men to adjust the ways that families are cared for have become a media event. Some progress has been made, but not as much as we had hoped, or as is often suggested. I believe that while we, as feminist educators and mathematicians, acknowledge this, our educational programmes often do not.

We try to keep our messages to girls simple and often they become simplistic. This may explain why many girls don't find our arguments particularly persuasive. Recently, colleagues and I interviewed a considerable number of 14-year-old girls and boys (Willis, 1991). With only a few exceptions, the girls saw their lives as tied to fundamental conflicts of choice. They were planning their futures and expected their lives as women to be segmented and manifold, while the boys were quite vague about their futures but perceived their lives more uni-dimensionally. Even amongst the 'post-feminist' girls who insisted that 'everything was equal now' and 'we can do whatever we like', there was almost always a 'but . . .'. In essence, these girls were telling us that they knew the gender-imposed constraints surrounding their choices were real, not imaginary. Yet we are quite often exasperated when we try to convince girls that they can be engineers and they persist in missing the point and asking who is going to tuck in the baby. When we tell them they can do anything, they suspect we mean that they are expected to do everything.

Fourth, approaches focusing upon 'changing choices' implicitly accept the hegemony of mathematics by which secure and high-status futures are seen as 'naturally' accruing to the mathematically privileged. As almost any adolescent will tell you, the true power of school mathematics lies not in its utility in further education and employment, but in its usefulness for accessing them. That is, a major function of mathematics is as a *de facto* intelligence test and consequently, as a filter between school and sometimes quite unrelated further educational and occupational opportunities. School mathematics, however, systematically disadvantages students not only on the basis of gender, but also on the bases of social class, race, and ethnic identity (Walkerdine, 1989; Dowling, 1991). Thus, school mathematics is deeply implicated in reproducing social inequality. The problem for feminists in mathematics education is that once having sold mathematics as an investment in the future, we can hardly attack the very exclusionary practices that maintain that investment potential.

Finally, interventions such as those I have been arguing against are typically marginal to the mainstream mathematics curriculum and fail to challenge it in any significant way. Indeed, they may actually reinforce the notion that 'the problems' of girls and mathematics lie outside the classroom and school. Over the past three years, some colleagues and I have undertaken a series of case studies in Australian schools to investigate the reception of gender-reform messages by students and teachers (Kenway and Willis, 1993). Given the high visibility of the 'problem of girls and mathematics', and a push from government and industry to increase the numbers of girls and women in mathematics, science and technology, we were not surprised to find that female participation in these subjects was a major topic of discussion in many schools. What was surprising, however, was just how little of an effect this had on the practices of mathematics education. Only a very few primary or secondary school teachers believed that the mathematics curriculum might be gender-inflected and that reform was either necessary or possible, for after all 'girls are as good as boys at mathematics and we don't make any distinctions between them.'

Several years ago, a television-advertising campaign designed to encourage girls to choose more and 'higher' mathematics was launched in several Australian states. Put simply, the message to girls was that 'mathematics multiplies your choices', and to parents that you should not 'pigeon-hole your daughter'. This campaign was read by many teachers we interviewed as confirmation that 'the problem' lies with the poor choices that girls make as a result of social and peer-group pressure, home influence, or a lack of accurate information about their options and futures. Thus, they appeared to hold the rather complacent belief that the curriculum was not an issue, and that it was equally appropriate 'for all students' in both content and pedagogy or, to the extent that it might not be, it was inherent in the 'nature of the subject' and, therefore, unalterable. Some suggested that the problem might be mathematics anxiety, low self-esteem, or a lack of confidence, but had 'no idea' how this might arise — these characteristics were, it seems, a corollary to being born female.

For these reasons, I believe that a focus on changing girls and/or their choices will not deliver justice in and through school mathematics. At best it will be insufficient to the task, and at worst may prove counter-productive. I suggest that for feminist mathematics educators, a more productive focus can be found in changing curricula.

Gender-Inclusive School Mathematics

Curriculum reforms designed to provide a different and better experience of mathematics for girls may, themselves, be informed by different feminist (and non-feminist) perspectives. As a result, they will vary in their extent and form. In this section, I will suggest some of the characteristics we might expect in a reformed mathematics curriculum.

Towards Non-sexist Curriculum Content

That school mathematics should not be sexist hardly needs to be said. There is some acceptance that we should use non-sexist language and illustrations. Certainly in Australia, many teachers would routinely expect textbooks to pass the 'non-sexist test'. That some of these changes are more apparent than real, and that much of the sexism of school mathematics is more deeply embedded than can be addressed by surface changes in text materials is perhaps less well understood or accepted.

Increasingly, we have come to understand how mathematics curricula tend to emphasize the experiences, concerns, and interests associated with masculinity rather than femininity. Clearly, success with mathematical tasks will come easier to students who already have a grounding with the ideas underlying the task, or who are familiar with the context in which the task is embedded. Heleen Verhage (1990) reported that she and her colleagues had been given the task of developing a mathematics curriculum which is both more realistic and more attractive to girls. By 'realistic mathematics education' is meant that the mathematics must be 'derived from the reality around us [as well as be] applicable to this reality' and furthermore, that students should 'rediscover (scraps of) mathematics and construct it themselves' (Verhage, 1990, p. 61). She observed a mathematics lesson on the study of symmetry and tessellation which drew on embroidery, a context more likely to be familiar to girls than to boys, and remarked,

> [I]t was my impression that the girls clearly felt stronger during this lesson. They were at an advantage right away as it took the boys a while to actually get started due, for instance, to having trouble threading the needle. The teacher told how, during the previous lesson, it had been just the other way around, when it concerned building brick walls. (Verhage, 1990, p. 66)

Examples such as this demonstrate the importance of context in enabling students to make the most of the learning opportunities provided in the mathematics classroom. Since many of the contexts in which mathematics is presented are more likely to centre in the lives of boys than girls, this gives boys a distinct advantage. Furthermore, such contexts must send quite clear messages to girls and young women about where mathematics came from, who it is for, and what its relevance is to them personally and to women collectively. Mathematics problems are likely to be either embedded in contexts more familiar and comfortable to boys, or decontextualized and 'de-peopled' altogether. Neither seems particularly friendly to girls.

Curriculum innovations such as that developed by Heleen Verhage seek to address this concern and are exciting and important. Unfortunately, the results of these efforts are often interpreted as simply adding to the repertoire of good activities and contexts from which particular mathematical concepts can be derived. Indeed, there is a widespread view that gender-inclusive school mathematics is largely a matter of making available the 'right' resource materials. As a teacher of mathematics said to me recently (private communication), 'I don't have any problems with that [gender equity in school mathematics] if the resources are there'. The 'right' resources are balanced, presenting both sexes 'equally', and providing a range of examples interesting to all. According to many teachers, curriculum writers, and publishers, the 'right' resources also take subtle or indirect approaches to gender equity. There are certain problems with this notion, to which I will return shortly.

Being Good at Mathematics

The work of Valerie Walkerdine and her colleagues has provided considerable insight into the gendered (and classist) construction of mathematical ability. Their early work (Adams and Walkerdine, 1986) suggested that children learn that success in mathematics requires a passive approach involving rote-learning and rule-following, while what is seen as intelligent behaviour in mathematics is active, exploratory, and rule-challenging; the former behaviour often being associated with femininity, the latter with masculinity. Girls, even high achievers in mathematics, are described as hard workers, plodders, capable, neat, conscientious, and helpful. Since these are not the characteristics which connote talent in mathematics, girls' early successes are not regarded as a sign of talent or even of real learning, 'instead of thinking properly, girls simply work hard' (Adams and Walkerdine, 1986, p. 84).

Walden and Walkerdine (1986) suggest that the contradiction between the practice of school mathematics which emphasizes rule-following, and the perception of the discipline that sees real mathematics learning as about challenging rules, leaves girls in an ambiguous position. It sets up a conflict between what is regarded as necessary to achieve femininity and be a 'good girl' by doing what the teacher asks, and what is necessary to be regarded as 'really good' at mathematics. Vicki Webber (1991), in a case study of one of her adult students named Jenny, provides a powerful account of the manner in which notions of being a 'good girl' are embedded in the practices of the school mathematics curriculum. Jenny's hold on mathematics is very fragile and she feels a fraud. As Webber (1991) remarks,

> When the knowledge comes from outside yourself, you don't own it, you have only a tenuous hold. You doubt the integrity of your own intellect . . . you are sure you will be found out . . . you hold your breath and wait to see if you've got it right. Maths is an extreme example of waiting for someone else to say, 'you've got it right, good girl'. (Webber, 1991)

Jenny drew parallels between learning to be Catholic and learning mathematics:

> Religious instruction, like maths, was filled with 'givens' and to question them was 'brazen, uppity'. Like the catechism she learned her maths 'by heart'. Like the Latin mass, maths was a secret language imbued with power, and which could only be understood by certain chosen initiates. Maths too was ancient wisdom that one could not question, could not presume to understand . . . [As Jenny says,] '. . . these theorems had more or less been handed down in the twelve tablets to Moses, and you've got no right to think about them because they've been developed by all these great minds. You've got nothing to contribute to this process. So you've got to learn them off by heart, to meet the requirements of being good, of being in the right.' (Ibid.)

For Jenny, 'maths was an altar with no steps up' and, in any case, girls could not be altar-boys. Webber recounts,

> Jenny felt in school that mathematics performance was closely associated with 'being a good thinker' and 'being mentally competent'. She had a strong sense in childhood that 'my mind wasn't good, but I can learn things by heart, and I can imitate quite well'. 'Real understanding' is the domain of the great minds of men, women's understanding comes from the path of faith and devoted effort. (Ibid.)

This is not to suggest that girls are forced out of mathematics by some sort of patriarchal conspiracy. Teachers of both sexes are as caught up in these processes as are students who, themselves, are not passive in the process as they struggle to make sense of conflicting messages. These contradictions, however, are part of a whole constellation of classroom practices which position males and females differently with respect to mathematics. As Walkerdine remarked, 'our education system in its most liberal form treats girls "as if" they were boys. The "clever girl" is positioned as though she could and can possess the phallus, while she has to negotiate other practices in which her femininity is what is validated.' (Walkerdine, 1990, p. 46) The achievement of gender equity in mathematics education will require us to address the ambiguity involved in school mathematics for girls, a formidable task, and one we have hardly begun.

Towards a Different Conception of Mathematics

While the research is still fairly limited, considerable anecdotal evidence suggests that girls and young women need to experience mathematics as a way of knowing and finding out, as well as to understand its tentative, intuitive, exploratory, human, and therefore fallible, nature. Jenny, for example, tells us that it can be different for her. After engaging in some problem-solving tasks, she said,

> After that I understood that this is how maths has developed through history. Mathematics is a shorthand for describing a real, meaningful process ... I knew what that formula was about. It was an expression of what I had actually done ... I know now where a formula comes from, that no matter how complicated it comes from real problems in the real world. (Webber, 1991)

It seems that our students need to know that mathematics makes sense, that *they* can make sense of it, and that they can work it out.

Several years ago, the Australian Government set up a programme which funded the development of 'gender-inclusive' teaching materials in mathematics and science. Mary Barnes was commissioned to produce a set of student text materials for teaching introductory calculus at the secondary level. As with the Netherlands project on realistic mathematics education, the goal was to model a curriculum that was non-sexist, used a more inclusive range of contexts, and acknowledged and incorporated women's contributions to both mathematics and society. It was to be rigorous and intellectually demanding, acknowledging and valuing some of the power of the 'disembeddedness' of mathematics, while at the same time embedding it in social concerns and people-oriented contexts. Furthermore, it was also to present mathematics as making 'human sense', non-arbitrary and fallible, produced by and for people, and having aesthetic, cultural, and instrumental importance. A rather ambitious agenda!

According to Barnes, the course stresses mathematical modelling and the solution of realistic problems, with less emphasis on traditional applications from the physical sciences and more emphasis on issues of importance in students' lives. It involves students working collaboratively to investigate problems, explaining and clarifying their thoughts in the process, while at the same time sharing insights, experience, knowledge, concerns, and confusions. The goal is for students to understand that mathematics should make sense and that they, themselves, are capable of deciding when something does or does not make sense.

The materials are now published (Barnes, 1991, 1993), and the indications regarding their use are encouraging. Although some students in the trials did not respond positively to the changes, most did. Overwhelmingly, their comments reflected an increased confidence in the quality of their knowledge, as well as a feeling of control over their own learning. They commented about understanding:

> **Girl** My dad said he never really understood what it was about ... I understand what it is about and I can do the problems whereas he can do the problems without really knowing.
> **Boy** When you get in groups you learn better, because you are mixing up your knowledge.
> **Boy** The course is much easier to understand because we know why.
> **Girl** They did make sense unlike other things we were doing in maths ... we were able to understand where they came from ... I enjoyed it.

about taking responsibility for their own learning:

> **Girl** At first the groups did what she [Mary, the author] said, you know had one of us writing and one timing ... But then only one person had the ideas written down, so if you looked at it at home you might not remember. So now we talk about it and decide what we all want to say and then we all write it down ... Sometimes we argue but not usually.
> **Girl** You have to make sure you stop and say in your group, Well, what's going to happen? and make sure everyone understands the why and then do it ... We just plugged it in [the computer] for a while and we didn't learn anything. But we realized.

and about purpose:

> **Girl** The best thing about them is how they related to real life. Most of the problems were practical. It makes it easier.
> **Boy** It's more real life so you had to think about it more. There were no real formulas.

One teacher, commenting about his female and male students, said, 'I have been so overwhelmed ... by the insight they have about graphs and the derivative function and what the derivative does ... They are just so much more powerful than the previous group. Their ability to justify things like the chain rule was breathtaking.' To be able to describe students as 'powerful' in doing mathematics is surely what we would like to achieve. Quite a few teachers commented on gender differences in response to the approach. Girls, they variously claimed,

- were more ready to share and to talk freely;

- were enjoying this work although they hadn't in the past;
- work better in groups;
- had more self-discipline;
- were more worried, but would come and see you about problems;
- were more creative;
- wanted to know why; and
- found the examples more interesting [than girls studying calculus have in the past].

Boys, by contrast, teachers said,

- were less ready to share their ideas;
- tended to waste more time;
- tended to be domineering, but now realize that other people have positive things to say;
- were more easily distracted; and
- were happy to just do it [rather than know why] and come up with the answer and think that's great.

It seems that in these classrooms, teachers and students of both sexes are beginning to change their perceptions about what mathematics is, who produces it and how, and what it means to learn mathematics and be good at it. Making such curricula a reality in schools is an ambitious goal, but were we to be successful, surely it would be both more rewarding and just. The fundamental question, however, is would it be emancipatory? Probably it would, but for it to fully reach its emancipatory potential, I believe it must be accompanied by explicit attention to gender and power relationships, as well as to how mathematics is implicated in perpetuating inequalities in these relationships.

Socially Critical Curriculum in Mathematics

Earlier I suggested that, to the extent that more 'gender-inclusive' curricula are developed, there is a preference for 'subtle' and 'indirect' approaches where the gender 'agenda' is not made explicit. As director of the programme through which Mary Barnes' materials were generated, I was constantly warned of the danger that, by being either too direct in our approaches, or too explicitly feminist, our efforts might be rejected by teachers. This dislike of direct approaches seemed to lie in their potential to be disruptive to the good progress of the class 'as a whole'. Potentially disruptive approaches included those which confront sexist assumptions and behaviours explicitly, and thus might offend some teachers, students or parents, or, perhaps, disturb or delay the achievement of the primary goal, that is, teaching 'the curriculum'. The argument was that nothing would be achieved if the materials were not used because teachers or students found them too threatening. Since the materials produced through the programme were intended for mainstream use, we succumbed and so few of our products were explicitly feminist. However, as illustrated by the following three examples, certain dangers are inherent in such indirect approaches.

Paul Dowling (1991) describes a well-known 'counter-sexist' mathematics textbook. Female names and pronouns are regularly used in conjunction with traditional male occupations and roles, and vice versa, but this is done without comment and the resulting dissonance is 'occasional and never explored' (Dowling, 1991, p. 3). Dowling suggests that this often becomes a target for humour, which may be counter-productive. He also recounts a story relating to an activity in the textbook which involves 'famous people', two being women, George Eliot and Florence Nightingale. The illustration was of a statue of George Eliot, with her name clearly printed below. One (female) student he spoke to assumed that the statue was of Florence Nightingale and had been misnamed. This student dealt quite readily with the dissonance involved in having a female figure given a 'male' name by explaining it to herself as an error. For this student, at least, the textbook was not at all disruptive.

Heleen Verhage (1990) recounts how a student in a lesson on symmetry and tessellation commented, 'This isn't math. This is sewing class', to which the teacher replied, 'When you have to draw a circle you don't call it drawing class, do you?' (Verhage, 1990, p. 66) The teacher's answer is valid, as is the underlying point that school mathematics is a social construction. Such occasional and incidental rebuttals, however, are unlikely to challenge the student's perception of what mathematics is, what counts as mathematics, and, therefore, what it means to be good at mathematics.

Students and teachers were interviewed during the trial period of Mary Barnes' calculus course. One common early reaction on the part of teachers was that the materials required more reading than students would do and that, in particular, this would disadvantage their weaker pupils. When asked to gather more information about who exactly was and wasn't doing the reading, and who was complaining, the teachers returned rather abashed. They had failed to notice that the original complaints had derived from a vocal minority of high-achieving boys, and a few high-achieving girls, who felt that there was more explanation provided than was necessary and it slowed them down. The students whom the teachers regarded as 'less able' insisted that they did the reading, and as one student fiercely demanded, 'Don't you remove a single word.' Moreover, it was predominantly the high-achieving male students who were most reluctant to participate in the group work that was an essential component of the course. These observations demonstrate how, unless we are vigilant, those already privileged in mathematics can continue to control and define what constitutes mathematics in school.

The above examples demonstrate that gender-inclusive materials do not by themselves ensure a gender-inclusive curriculum. In each case, interpretations were based on existing frameworks and, in the absence of any disruption, the potential for significant change was not there. I would argue that the effect of the intervention would be more powerful in each instance if the agenda were made clearer. What is needed is a more socially critical mathematics curriculum in which mathematics is regarded both as a powerful tool for critique, and also as an object of critique. Allow me to provide just a glimpse of what this might mean.

Women have traditionally sold their skills cheaply because they and others either do not recognize them, or may even denigrate them (see, for example, Pocock, 1988; Holland, 1991). Often, women have not been prepared to believe that they had any skills worth possessing unless they were male-defined. This is as true of their mathematical labour as it is of their labour in general. As Mary Harris (1989; see also this volume, Chapter 9) has pointed out, designing the lagging for a pipe that turns at right angles is an engineering feat requiring significant mathematics, while designing a sock for the heel of a foot is a trivial knitting task requiring insignificant mathematics. However, as I have already suggested, simply ensuring that socks and pipes, embroidery and bricks are equally represented in a mathematics curriculum is unlikely to be enough. The question of why sewing isn't mathematics, while bricks and circle drawing are, needs to be addressed with students. Indeed, Heleen Verhage (1990) raised the vexing question of whether activities such as embroidery are emancipatory or merely serve to confirm traditional role patterns. In my view, the reception and effect of such activities is likely to be influenced by the extent to which their gendered nature is made explicit. As a colleague and I have argued elsewhere (Kenway and Willis, 1993), 'home' still signifies the place of the female, apart from the public and important world of the male. By leaving 'home' out, we may actually emphasize its unimportance and hence female unimportance, thus reinforcing rather than reducing gender stereotypes. As I suggested earlier, to be equal is to be male.

A recent Australian newspaper advertisement promoting a city building project which would generate 'thousands of jobs' provided a series of about fifty male and female figures labelled with the name of a job and the number of positions to be generated if the project went ahead. In the advertisement, 8.3 per cent of the total number of jobs were represented by female figures, and not a single one was of non-European origin. Women and minority Australians remain the silenced unemployed. A close look at the poses and style of dress of the figures also revealed appalling gender, class, and racial stereotypes. Images, such as these, do the most damage when their sexism, classism, and racism go unnoticed at the conscious level and are thus absorbed as natural or accurate descriptions of the world. The mathematics curriculum can make students aware of such things in order to deal with them. Issues relating to unemployment and 'women's work', both paid and unpaid, might become the focus of any one of a number of short or extended projects in mathematics. We might, for example, ask students to use their mathematics to investigate the consequences on women, both individually and collectively, as well as on the entire community, of not defining domestic labour as 'work'.

Elsewhere (Willis, 1994), I have described an Australian series of mathematics textbooks which on the surface appeared to have made considerable effort to be non-sexist by using gender-neutral language and illustrations, and by having girls and boys equally represented and equally and actively involved in 'real-world' mathematics. However, an analysis of the representations of adults in the texts provided quite a different picture.

> Almost all women (except teachers) are referred to as mum or grandma or the butcher's wife, almost always cooking or shopping, and rarely given a name. Men, on the other hand, are occasionally called dad ... but are mostly given names and described in terms of their job. Men have and handle more money. To what extent do these books help to convince boys and girls that their mothers should be at home and available to help out at school, and their fathers should be supporting the family financially? And do they imply that girls and boys should chose to emulate these more 'suitable' mothers and fathers? While mathematics is now regarded as important for all students, perhaps lower levels of mathematics are sufficient for women's work. (Willis, 1994)

One approach to dealing with such materials might be not to use them. Is it conceivable, instead, that we might use them as an object of study, that students might be asked to investigate how men and women are constructed in their mathematics textbooks? Might not they also use their data collection and analysis skills to investigate who controls the classroom agenda and whose interests are met by the materials they use?

I commented earlier on an Australian television campaign aimed at encouraging more girls to participate in mathematics during their upper-secondary years. The advertisements asserted that you have 400 per cent more choices if you do mathematics. Unsuccessfully, I tried very hard to find justification for such figures and, indeed, the explanation of the data which was provided to me showed a clear misuse of mathematics. Were we seriously claiming that 80 per cent of all jobs, of every kind of job for every kind of girl, needed mathematics in the final two years of schooling? Of course not. By increasing participation in mathematics, will we be increasing the number of jobs available? Again, the answer is no. Most popular analyses I have read which attempt to forecast where the jobs will be in the future suggest that mathematics will not be central in doing them, although it may well be crucial for getting them. Nevertheless, the advertisements were 'successful' according to the advertising company's criteria, that is, there was good 'saturation', a lot of people remembered having seen them. But what, I wonder, was the effect of the campaign on the morale of those students whom school mathematics has not served well? And what was the effect on young women and men of the portrayal in the advertisement of girls in a foetal curl squashed into boxes looking sad and oppressed ('don't pigeonhole your daughter')? Had these images of young women appeared in a men's magazine, we would have been scandalized.

To those feminists who were critical of these campaigns, the only winners in this campaign were the employers, for they would now have more choices. Several colleagues tried to have the 'voice over' changed before the advertisement was broadcast to remove the reference to '400 per cent' on the grounds that it was distorting and misleading. They were told that the advertisement would lack 'punch' without the numbers, since 'people find figures very persuasive'. Thus, the campaign itself provided an example of the use of mathematics to distort, intimidate, and mystify. Mathematics was used *on* girls, not *for* them. This is precisely the sort of media claim that we would hope a quality mathematics education could enable students to critique. A productive use of this advertisement would be to encourage students to deconstruct it. Students should be able to interrogate the claims, asking what the 400 per cent actually refers to and how it was determined, thus evaluating the sexist and classist assumptions upon which the advertisement was based. This task would certainly not be beyond the capabilities of most 15-year-olds and has enormous emancipatory potential. Of course, the inevitable questions which would follow are: What is it about mathematics that encourages people to suspend their common sense and hence allow such material to go unchallenged and whose interests are served when such programmes 'go to air'? The difficulty in adopting such approaches is in finding the balance between being too disruptive and not being disruptive enough. In the former case, students may feel so uncomfortable that they resist completely and are unable to engage. In the latter case, they may comfortably accommodate new ideas within their existing frameworks and thus find no need to change their thinking.

Conclusion

The task of mathematics education should, explicitly and unapologetically, be the provision of a rewarding, just, and emancipatory educational experience for girls. And the justification for doing so is precisely that it is rewarding, just, and emancipatory. However, this should not come about by providing alternative curricula for girls. Everyone's attitudes and beliefs about what constitutes mathematics, what counts as valued knowledge in mathematics, and what it means to be good at mathematics must change. Consequently, students of both sexes will need to experience mathematics differently than they do currently.

Mathematics curricula should aim to enhance female life options by making choices more fully and equally available. Girls and young women need to understand how their attributions and choices about mathematics, as about all else, are socially constructed and constrained. A gender-expansive and more critical mathematics curriculum would address the ways in which mathematics is used to intimidate and mystify, thus confronting the issue of gender formation in school explicitly. This, in turn, would enable both girls and boys to understand how they are positioned differently with respect to mathematics, to identify the processes by which gendered patterns of achievement and participation in mathematics are produced and naturalized, and to recognize the effect of these processes on their futures. Over a decade ago, Jean Blackburn, a former Australian Educator of the Year who first used the term 'sexually inclusive curriculum' in the early 1980s, suggested that one role of education is to help young people 'to understand and reflect on what is, on how it came to be that way and on what they want to do about it, both at the level of their own lives and in relation to social action' (Blackburn, 1982, p. 10). The purpose of studying gender relations within a socially critical framework is not to point and move girls (or boys) in certain directions, but rather to widen the range of positions available to them both now and in the future.

Note

1 These points were developed in collaboration with Jane Kenway, Deakin University, Australia and the Girls Unit of the Education Department of South Australia.

References

ADAMS, C. and WALKERDINE, V. (1986) *Investigating Gender in the Primary School*, London, Inner London Education Authority.

BARNES, M. (1991, 1993) *Investigating Change: An Introduction to Calculus for Australian Schools*, Units 1–10 and Teachers' Handbooks, Carlton, Victoria, Curriculum Corporation.

BLACKBURN, J. (1982) 'Some dilemmas in non-sexist education', *The Secondary Teacher*, 1, pp. 10–11.

DOWLING, P. (1991) 'Gender, class, and subjectivity in mathematics: A critique of Humpty Dumpty', *For the Learning of Mathematics*, 11, 1, pp. 2–8.

HARRIS, M. (1989) 'Textiles and maths', *GEMS*, 1, 4, pp. 37–9.

HOLLAND, J. (1991) 'The gendering of work', in HARRIS, M. (Ed) *Schools, Mathematics and Work*, London, Falmer Press, pp. 230–52.

JOHNSON, R. (1983) 'Educational politics: The old and the new', in WOLPE, A. and DONALD, J. (Eds) *Is There Anyone Here from Education?*, London, Pluto Press, pp. 11–26.

KENWAY, J. and WILLIS, S. (1993) *Telling Tales: Girls and Schools Changing their Ways*, Canberra, Department of Employment, Education and Training.

KENWAY, J., WILLIS, S. and JUNOR, A. (1994) *Critical Visions: Curriculum and Policy Rewriting the Future*, Canberra, Department of Employment, Education and Training.

POCOCK, B. (1988) *Demanding Skill: Women and Technical Education in Australia*, Sydney, Allen and Unwin.

VERHAGE, H. (1990) 'Curriculum development and gender', in BURTON, L. (Ed) *Gender and Mathematics: An International Perspective*, London, Cassell, pp. 60–71.

WALDEN, R. and WALKERDINE, V. (1986) 'Characteristics, views and relationships in the classroom', in BURTON, L. (Ed) *Girls into Maths Can Go*, London, Holt, Rinehart and Winston, pp. 122–46.

WALKERDINE, V. (1989) *Counting Girls Out*, London, Virago Press.

WALKERDINE, V. (1990) *Schoolgirl Fictions*, London, Verso.

WEBBER, V. (1991) 'Dismantling the Altar of Mathematics', Unpublished paper.

WILLIS, S. (1994) 'Mathematics: From constructing privilege to deconstructing myths', in GASKELL, J. and WILLINSKY, J. (Eds) *Gender In/forms Curriculum: From Enrichment to Transformation*, New York, Teachers College Press.

Chapter 23

We Don't Even Want to Play: Classroom Strategies and Curriculum which Benefit Girls

Joanna Higgins

Introduction

Mathematics games, puzzles, and constructions are commonly used along with other independent activities in the classroom to familiarize students with particular mathematical concepts. The number of players, the amount of time spent participating, and the range of skills and concepts explored will vary in any one activity. In this chapter, I will discuss the ways in which boys and girls interact during these independent activities. I will highlight the factors which contribute to a differential experience between girls and boys, in which the former are disadvantaged as mathematics learners. I will suggest that the manner in which males respond to the teacher's chosen curriculum shapes the learning experience. My analysis of interactions between children playing mathematics games will show how girls are disadvantaged in this context. Finally, I will propose classroom strategies and curriculum which are more likely to benefit girls.

Gendered Achievement and Order in the Classroom

Clark (1990, p. 2) has pointed out that 'the primary [school] years are crucial in the construction of gendered achievement and aspirations and that the neglect of these years in most gender equity programmes is a serious problem which must be redressed'. When trying to identify micro-processes operating within the classroom that offer potential changes to benefit girls, it is important to make the gender dynamics of the classroom visible for teachers and educators. Alton-Lee and Densem (1992), for example, have drawn attention to phenomena that are often hidden from both teachers and classroom observers. They suggest that the existence of subtle gender and racial bias in the classroom requires the debate about gender-inclusive curriculum to move beyond issues of access and participation. More specifically, it needs to contend with the hidden processes of classroom structures, as well as examine the differences between the intended, the implemented, and the perceived curricula.

Power structures of the classroom influence student behaviour as well as shape the learning environment in ways which benefit boys more than girls. The teacher's predominant concern is with maintaining order to facilitate complex, autonomous, learning activities. Fennema and Peterson (1985) identify interactions with the teacher as key factors in the development of autonomous learning behaviours necessary for high-level cognitive tasks such as those in mathematics. They note that the teacher is less likely to direct girls back to a task because their behaviour is often seen as less disruptive than that of boys. Furthermore, males receive more disciplinary contacts, and more praise, when on-task. This suggests that boys' learning is facilitated by the teacher's concern with maintenance of order. Such structures in the classroom give boys an advantage in the learning situation. An important factor which requires consideration is the manner in which girls perceive the teacher's actions. The fact that the teacher's attention is directed away from the girls' quiet learning to boys' off-task behaviour simply reinforces the male advantage. Girls, as a result, arrive at the belief that their learning is less important.

Freedom of Choice

The provision of choice has often been championed by teachers as a means of meeting the individual needs of learners. In practice, choice refers to the provision of a range of options which relate to a set of learning goals and from which children are expected to select. However, the way the teacher structures the learning environment constrains this freedom of choice. Similarly, learning outcomes are influenced not only by teacher assumptions about intended learning, but also by gender dynamics which are often not taken into account. Many commentators have argued for a reconsideration of the child-centred model on which such classroom practices are based. Clark (1990, p. 26) points out that 'child-centred practices can allow teachers to avoid facing this responsibility [for providing equitable learning opportunities] by allowing teachers to believe that children can "freely" choose and that we can be neutral'. The provision of freedom of choice needs to be re-examined, not only for the reasons I have already stated, but also because boys' choices are commonly treated as being of more value. This results in boys 'owning' certain types of activity.

Having Access

Boys' ownership of certain activities is illustrated in the following transcript of a classroom segment in which girls are simply given access to the blocks.[1] Gender dynamics are evident during the construction of a three-dimensional model of a building. The boys' possession of this activity proves to be too much for the girls in the end, and the learning opportunity intended by the teacher is sabotaged. Rather than persevering in their own model building goals, the girls learn to succumb to the boys' demands.

> Two groups are working at the same activity — the blocks. One is a group of boys, the other a group of girls. The boys need more blocks to make their model bigger. While searching around for these, they look across to the girls' model.
>
> **Michael:** We need a block like you've got.
>
> He jumps across to the model and takes one. The girls try unsuccessfully to stop him and this leaves a gap in their model.
>
> **Shaun:** We might build a huge place and we might use up all the blocks.
>
> **Martha:** No, Shaun.
>
> Shaun threatens to grab some more of the girls' blocks.
>
> **Martha:** No Shaun, we might bash.
>
> An observer asks Philip to tell her about the model that the boys have made.
>
> **Philip:** It's a big castle just for people and people can go up there, [pointing], and down there, over there on the top to look out for some gold.
>
> She comments that she heard him asking the girls if they would like to join.
>
> **Philip:** We were going to join them, but they didn't let us.
>
> He points to another part of the model.
>
> **Philip:** And cars who have broken down would get fixed here.
>
> She asks him why they wanted to join up with the girls.
>
> **Philip:** We wanted to join up so it could be really really big.
>
> Terry comes over so she asks him to tell her about the model.
>
> **Terry:** I made a big building so you could jump across the sea.
>
> The observer moves across to the girls' model and talks with Martha.
>
> **Martha:** It's a school. These are the seats [pointing]. That's the playhouse. That says 'school today'.
>
> Philip comes over to the girls' model.
>
> **Martha:** Would you leave it, Philip.
>
> The girls' model is fast disappearing. The observer asks all the girls why they didn't want to join up with the boys.
>
> **Martha:** Just didn't want to.
>
> **Nadine:** They always say if you try and bust up ours we'll bust up yours.
>
> Michael comes over to get some more blocks. The girls finally give into his demands.
>
> **Nadine:** You can have these blocks if you want 'cos we don't even want to play with them.

It is evident that gender dynamics intervene not only in the structure of this learning opportunity, but also in the intended learning outcomes. The boys actively engage in model building and achieve their goal of making 'the biggest'. The girls, on the other hand, learn only to give up their own goal in the face of boys' aggressive demands. This example suggests that the notion of freedom of choice as a means of achieving equity in learning is an illusion. Gender dynamics will commonly intervene in such experiences, and teachers must adjust for this by devising techniques that facilitate girls' learning.

Being the Teacher

Walden and Walkerdine (1986) have argued that issues of dominance and subordination can be seen in the social relations of the classroom. For example, Bird (1992) found that girls seem to be much more willing than boys to assume the role of teacher. In my own observations, there was an obvious difference in the nature of the leadership roles taken up by girls and boys.

> Four children are participating in a board game as part of their independent mathematics activities. It is designed to help children learn how to order in numerical sequence, as well as understand the 'one more than' relationship.
>
> **Anne:** Whose turn?
> **Mary:** Martha's.
> **Tim:** Mine.
> **Anne:** Tim.
> **Martha:** You had lots of turns.
>
> Tim takes his turn.
>
> **Anne:** Go Martha. One more to go to win.
>
> Martha has her turn.
>
> **Tim:** I've got two more to go.
> **Anne:** Your turn.
>
> Mary plays.
>
> **Martha:** My turn.
> **Mary:** Yes. Four.
> **Anne:** Martha, don't cheat. It was Tim. Martha your turn. Right now Tim.
> **Martha:** Right your turn.
> **Tim:** Supposed to be my turn.
>
> Tim knocks his card off the board.
>
> **Anne:** Don't Tim. Hey cheat. [to Mary]
> **Martha:** Your turn, Timmy.
> **Tim:** My turn.
>
> Anne mixes up the cards.
>
> **Anne:** Now you can't get two fish to go there. Tim, your turn.
>
> Tim has his turn.

In this game Anne took on the role of the teacher. She took control of the game by directing the turns and by encouraging players not to abandon hope when they seemed to be losing. She also monitored the game for cheating, and when the only boy in the game misbehaved, she reprimanded him. Clark (1990) has suggested that boys and girls have available to them different ways of being powerful. Boys have access to direct forms of power such as sexual harassment or direct challenges to the authority of the teacher. Girls are more likely to gain power through indirect means, for instance, by being a quiet cooperative model pupil, or by taking on the role of teacher.

Commentators such as Clarricoates (1987) have suggested that girls avoid risks in their learning by adopting feminine roles. In this context, assuming the role of the teacher should be interpreted as the child conforming to the relatively submissive expectation of 'helpful pupil'. Encouraging girls to behave in this fashion merely sustains them in the traditional feminine position of subordination. However, arguing that 'things are not always what they seem', Bird (1992) has suggested that the girl who plays 'teacher' is instead taking up a powerful position by claiming the authority of the teacher. Similarly, Walden and Walkerdine (1986, p. 125) suggest that 'by being positioned like the teacher and sharing her authority, girls are enabled to be both feminine and clever; it gives them considerable kudos and helps their attainment'. As teachers, we need to critically examine the social content of the roles we provide for girls in a mathematics learning activity. We need to ensure that girls are adopting roles which will enhance their learning of mathematics. The creation of all-girls groups will not necessarily achieve this on its own.

Girls and Competition

Children's own gendered expectations about game-playing behaviours have a substantial impact on participation patterns. In particular, it is difficult for some girls to respond to competitive situations in the classroom. The two transcripts which follow demonstrate the contrast in gender-related reactions to a competitive environment. The activity is an adaptation of the board and card game 'Memory' which has been designed to help children develop a concept of numerical order. 5-year-old children, in groups of four, attempt to be the first to cover all the marked spaces on their board with a card. A space can be covered if a child turns over a card from those spread out face down which shows a set containing one more element than the set shown on the corresponding space of their board. While they engage in this undertaking, the teacher is involved in an instructional activity with another group.

Philip, Terry, Shaun and Michael are sitting playing a board and card game. Philip takes his turn.

Philip: You've got four points. I've got four points too. Both got four points.

Terry has his turn, but conceals his card instead of showing it to the others as is normally done in 'Memory'. One aspect of the game is remembering where particular cards are positioned among all the cards when they are placed face down. None of the other players object to his cheating.

Philip: Oh, you got five points too.

Terry: Have you seen any of mine?

Terry tells an observer that you have to sneak a look at the cards.

Terry: Do you have six points?

Philip counts.

Philip: Yeah, I got six points.

Terry: Who's won?

Philip: I got six. Let's play again. Now people, help me turn them over.

The boys then start the next round of their game.

The above scenario documents the manner in which the competitive aspect of this game with its goal-orientation is compatible with the boys' learning style. They appear to be motivated by the collecting of points and the chance to win the game, as shown by their enthusiastic and immediate playing of the next round. For these reasons, teachers often use competitive activities to minimize the potential for boys' disruptive behaviour.

We see a marked contrast to this in the responses of a group of girls playing the same game.

Anne: Martha, your turn.

Martha: You had one in your hand.

Anne: Don't let Mary see, Martha.

Martha: I won, I won, I won.

Anne: You lose, Mary. Three more turns. I won after Martha.

Martha: I won first.

At this point Mary starts crying.

Mary: You're cheating.

Martha gets up at once to comfort Mary by putting her arm around her.

Martha: I didn't say anything.

Tam: It's only a game.

Mary is still crying loudly. At this point the teacher comes over.

Mrs S: Oh Mary.

Mary tells her that the others cheated.

Mrs S: I think you and Anne should play somewhere else as you were nearly in tears the other day as well.

The girls are concerned both with monitoring the procedures of the game, and with supporting all members of the group. When one participant becomes upset, the others comfort her and are diverted from continuing with the game.

Participation Patterns

When we compare the respective responses of each gender group to this game, we see that there are differences in attitudes, interactions, and participation patterns, and that these all influence learning outcomes. The words chosen by each group serve as a window to the attitudes involved in the playing of the mathematics game. The boys appear to have a much more competitive attitude, indicating their concern with winning as the primary goal of their interaction. Most of their conversation revolves around how many 'points' each has accumulated. The girls, by contrast, talk of 'turns'. This indicates their preoccupation with the procedures and process of the game, rather than with the goal of winning. Thus, while the boys learn to win, the girls learn to play the game. It is interesting to note that in all of the activities described here, learning the mathematical concepts involved is almost incidental.

An additional aspect of gender differences in game-playing can be witnessed in the way each group responds to cheating. In the female group, Anne in her 'teacher role' instructs Martha not to let Mary see the cards. This is followed by Mary's accusation that Martha is cheating when Martha claims she has won. In contrast, Terry's query about others having seen his cards, as well as his assertion that it is necessary to sneak a look at the cards of other players, indicate that the boys harbour the belief that cheating in order to win is a necessary part of playing games such as 'Memory'.

The overall participation patterns of the two groups can also be seen to differ. The interactions among the boys appear to be less structured in terms of rules and turn-taking than that among the girls. Consequently, the boys are able to achieve the goal of rapidly placing cards in order to show the one-more-than relationship. The girls, however, seem to lose sight of this goal altogether. In their group, the structuring of the game procedures is provided through someone adopting the 'teacher role', while in the boys' group this does not occur. The boys appear to engage in self-monitoring of their progress as they work towards the goal of earning six points. On the other hand, the girls seem to be more concerned with monitoring each other's behaviour and conformity to game procedures. As a result, learning outcomes for the girls are primarily social rather than mathematical in nature. In sum, since gender differences in learning styles influence learning outcomes, girls may fail to engage with the intended mathematical concepts in competitive game situations.

Conclusions

It is imperative for teachers to realize that gender expectations and gender dynamics may intervene in the intended learning experience. In this way, girls' learning can be severely disadvantaged. As has been illustrated in the preceding discussion, the importance conveyed upon the maintenance of order in the classroom has several unintended results. One is that boys' learning styles and dominance are reinforced through the provision of competitive activities designed to minimize their disruptive behaviour. Another is that because they are not as disruptive in the classroom, females receive less teacher attention and constructive concern than their male counterparts. On a related note, females are furnished with less encouragement to stay with the learning task. Taken together, the impact on female learning is very serious.

All of this suggests that if girls are to benefit from independent mathematics tasks, more cooperative, as opposed to competitive, activities need to be offered. While only a first step, such an approach would better match girls' observed learning styles. A more carefully selected range of choices should function to help girls find activities in which they can achieve success. Furthermore, the provision of a sufficient quantity of equipment would reduce the possibility of girls being 'squeezed out' by boys. In this way, boys 'ownership' of particular activities could be diminished.

It may be necessary for teachers to reconsider the assumptions on which the provision of teacher-independent exercises are based. As the preceding examples show, gender dynamics will often intervene in activities which are grounded in the child-centred principle of choice. Neither access, nor opportunity, will be equal unless teachers restructure the learning situation to support the needs of both girls and boys. Girls need to be encouraged not only to persevere in problem-solving, but also to engage in risk-taking behaviour.

In order to support all learners, teachers need to monitor the group interactions which take place during independent mathematics activities and to manage gender dynamics. An important means of such support is the teaching of social skills that are crucial for learning and necessary for engaging in cooperative activity. The skills for working in groups, for listening to others, for generating ideas collaboratively, and for participating in reflective discussion cannot be assumed inherent in young children. Furthermore, these skills are critical in the fostering of mathematical development. Designing a learning environment in which all students engage fully with particular mathematical concepts requires at least as much attention to gender dynamics as to mathematics itself. This is particularly true when independent activities are used for pedagogical purposes.

Note

1 Observations cited in this chapter are drawn from research for my masters thesis (Higgins, 1991). Data was gathered over a thirty-week period during 1989 using standard ethnographic practices.

References

Alton-Lee, A. and Densem, P. (1992) 'Towards a gender-inclusive school curriculum: Changing educational practice', in Middleton, S. and Jones, A. (Eds) *Women and*

Education in Aotearoa 2, Wellington, Bridget Williams Books Ltd., pp. 197–220.

BIRD, L. (1992) 'Girls taking positions of authority at primary school', in MIDDLETON, S. and JONES, A. (Eds) *Women and Education in Aotearoa 2*, Wellington, Bridget Williams Books Ltd., pp. 149–68.

CLARRICOATES, K. (1987) 'Child culture at school — A clash between gendered worlds?' in POLLARD, A. (Ed) *Children and Their Primary Schools*, London, Falmer Press, pp. 188–206.

CLARK, M. (1990) *The Great Gender Divide. Gender in the Primary School*, Melbourne, Curriculum Corporation.

DAVIES, B. (1988) *Gender, Equity and Early Childhood*, Commonwealth of Australia, Curriculum Development Centre.

FENNEMA, E. and PETERSON, P. (1985) 'Autonomous learning behaviour: A possible explanation of sex-related differences in mathematics', *Educational Studies in Mathematics*, 16, 3, pp. 303–20.

HIGGINS, J. (1991) 'Mungapungas and Brooming Revealed: Independent Learning in Early Mathematics', Unpublished masters thesis, Victoria, University of Wellington.

WALDEN, R. and WALKERDINE, V. (1986) 'Characteristics, views and relationships in the classroom', in BURTON, L. (Ed) *Girls Into Maths Can Go*, London, Holt, Rinehart and Winston, pp. 122–46.

Chapter 24

Moving Towards a Feminist Epistemology of Mathematics[1]

Leone Burton

Introduction

Received science has been criticized on three grounds from a gender perspective. The first is its reductionism and its claim to be objective and value-free (for example, Harding, 1986, 1991; Keller, 1985; Rose and Rose, 1980). Second, the conventional style of learning and teaching in science, its pedagogy, has been challenged. It is suggested that enquiry methods used by scientists are often intrusive and mechanistic, separating observer and observed, and reinforcing competition. Further, these methods are presented not only as 'correct' but also as the only way possible (for example, Kelly, 1987; Whyte *et al.*, 1985). Third, having rejected objectivity as an untenable criterion for judging science, a new scientific epistemology was required and has been derived (see Rosser, 1990) by examining the connections between the discipline and those who use it, and the society within which it develops. This line of reasoning is consistent with a broad range of thinking in the sociology of science.

> The old certainties about science, the old belief in its cultural uniqueness and the old landmarks of sociological interpretation have all gone. (Preface by Michael Mulkay to Brannigan, 1981, p. vii)

Mathematics and mathematics education have been subject to a similar challenge from within on philosophical, pedagogic and epistemological grounds. The philosophical arguments for a rejection of absolutism in mathematics have been explored elsewhere (see Ernest, 1991). Lakatos (1976, 1983), Bloor (1976, 1991) and Davis and Hersh (1983) have all made similar philosophical and epistemological criticisms to those outlined in the science literature with respect to the so-called objectivity of mathematics.

Likewise, a critique similar to that found in science has been made of mathematical pedagogy (see, for example, Burton, 1986, 1990a; Fennema and Leder, 1990; Leder and Sampson, 1989). Despite 'many reports calling for curriculum reform in mathematics and science ... the reforms suggested do not take feminist concerns into account; in fact, in the case of mathematics they tend to put added emphasis on curricular areas in which young women regularly perform less well than their male counterparts' (Damarin, 1991, p. 108). Mathematics tends to be taught with a heavy reliance upon written texts which removes its conjectural nature, presenting it as inert information which should not be questioned. Predominant patterns of teaching focus on the individual learner and induce competition between learners. Language is pre-digested in the text, assuming that meaning is communicated and is non-negotiable. In Robert Hull's terms this defines 'knowledge as an object and so equates knowing, and coming to know, with its possession; it effaces the crucial distinction between the learner's subjective experience of moving towards knowledge and the objectifying of a knowledge finally achieved' (Hull, 1985, pp. 49–50).

Like science, therefore, mathematics is perceived by many students and some teachers as 'a body of established knowledge accessible only to a few extraordinary individuals' (Rosser, 1990, p. 89). Indeed, the supposed 'objectivity' of the discipline, a cause for questioning and concern by some of those within it, is often perceived by non-mathematician curriculum theorists as inevitable (see, for example, Hirst, 1965, 1974, and, for a critique expanding the points being made here, Kelly, 1986). But 'the processes of knowing (and so also of science) in no way resemble an impersonal achievement of detached objectivity. They are rooted throughout ... in personal acts of tacit integration. They are not grounded on explicit operations of logic. Scientific inquiry is accordingly a dynamic exercise of the imagination and is rooted in commitments and beliefs about the nature of things' (Polanyi and Prosch, 1975, p. 63). Adopting an objectivist stance within mathematical philosophy means accepting that mathematical 'truths' exist and the purpose of education is to convey them into the heads of the learners. This leads to conflicts both in the understanding of what constitutes knowing, and of how that knowing is to be achieved through didactic situations. For example, such conflicts can be found between the UK mathematics national curriculum, expressed in terms of a hierarchy of mathematical truth statements, and the support documentation given to teachers which includes such relativistic statements as:

> Each person's 'map' of the network and of the pathways connecting different mathematical ideas is different, thus people understand mathematics in different ways. (Non-Statutory Guidance to the Mathematics National Curriculum, par. 2.1, p. C1)

> The teacher's job is to organise and provide the sorts of experiences which enable pupils to construct and develop their own understanding of mathematics, rather than simply communicate the ways in which they themselves understand the subject. (Ibid., par. 2.2, p. C2)

Although 'the ideal of pure objectivity in knowing and in science has been shown to be a myth' (Polanyi and Prosch, 1975, p. 63), it is a philosophical myth which continues to exercise enormous power over mathematics both in curricular and in methodological terms.

Proposed as an alternative, social constructivism is a philosophical position which emphasizes the interaction between individuals, society and knowledge out of which mathematical meaning is created. It has profound implications for pedagogy. Classroom behaviours, forms of organization, and roles, rights and responsibilities have to be re-thought in a classroom which places the learner, rather than the knowledge, at the centre. Epistemology, too, requires reconsideration from a theoretical position of knowledge as given, as absolute, to a theory of knowledge, or perhaps better, of knowing, as subjectively contextualized and within which meaning is negotiated.

With respect to science, Sue Rosser (1990) has stated:

> If science is socially constructed, then attracting a more heterogeneous group of scientists would result in different questions being asked, approaches and experimental subjects used, and theories and conclusions drawn from the data. (Rosser, 1990, p. 33)

How might including many of those currently outside the mainstream of mathematical development influence its conjectures, its methods of enquiry and the interpretation of its results? In turn, how might any changes which resulted from a philosophical shift, affect the pedagogy and epistemology of the discipline? In particular, what are the epistemological questions which are sharpened by bringing a feminist critique to bear on the discipline of mathematics? These are the focus of this chapter.

Adopting a Cultural View of Mathematics

In writing about mathematics, Sandra Harding (1986) drew attention to its cultural dependency:

> Physics and chemistry, mathematics and logic, bear the fingerprints of their distinctive cultural creators no less than do anthropology and history. A maximally objective science, natural or social, will be one that includes a self-conscious and critical examination of the relationship between the social experience of its creators and kinds of cognitive structures favoured in this inquiry . . . whatever the moral and political values and interests responsible for selecting problems, theories, methods, and interpretations of research, they reappear at the other end of the inquiry as the moral and political universe that science projects as natural and thereby helps to legitimate. (Harding, 1986, pp. 250–1)

Despite the stance taken by many mathematicians on the objectivity and value-free nature of the discipline, David Bloor convincingly argued from a historical perspective that it is possible to conceive of alternative mathematics differently derived at different periods:

> Seeing how people decide what is inside or outside mathematics is part of the problem confronting the sociology of knowledge, and the

> alternative ways of doing this constitute alternative conceptions of mathematics. The boundary (between mathematics and meta-mathematics) cannot just be taken for granted in the way that the critics do. One of the reasons why there appears to be no alternative to our mathematics is because we routinely disallow it. We push the possibility aside, rendering it invisible or defining it as error or as non-mathematics. (Bloor, 1991, pp. 179–80)

More recently, Sandra Harding has pushed the argument further to locate mathematics firmly within its interpretative context despite its overtly comparable formalistic expression:

> There can appear to be no social values in results of research that are expressed in formal symbols; however, formalization does not guarantee the absence of social values. For one thing, historians have argued that the history of mathematics and logic is not merely an external history about who discovered what when. They claim that the general social interests and preoccupations of a culture can appear in the forms of quantification and logic that its mathematics uses. Distinguished mathematicians have concluded that the ultimate test of the adequacy of mathematics is a pragmatic one: does it work to do what it was intended to do? Moreover, formal statements require interpretation in order to be meaningful ... Without decisions about their referents and meanings, they cannot be used to make predictions, for example, or to stimulate future research. (Harding, 1991, p. 84)

In his discussion of mathematical epistemology, George Joseph (in Nelson *et al.*, 1993) drew attention to two major philosophical presuppositions which underlie western (European) mathematics. These are, first, that mathematics is a body of absolute truths which are, second, argued (or 'proved') within a formal, deductive system. However, he pointed out that dependence upon an axiomatically deduced system of proof was a late nineteenth-century development which was pre-dated by 'proofs' closer in style to that of non-European mathematicians:

> The Indian (or, for that matter, the Chinese) epistemological position on the nature of mathematics is very different. The aim is not to build up an imposing edifice on a few self-evident axioms but to validate a result by any method, including visual demonstration. (Nelson *et al.*, 1993, p. 9)

George Joseph further stated that:

> None of the major schools of Western thought ... gives a satisfactory account of what indeed is the nature of objects (such as numbers) and how they are related to (other) objects in everyday life. It is an arguable point ... that the Indian view of such objects ... may lead to some interesting insights on the nature of mathematical knowledge and its validation. Irrespective of whether this point can be substantiated or not, a more balanced discussion of different epistemological approaches to mathe-

> matics would be invaluable. However, a different insight into some of the foundational aspects of the subject is hindered by the prevalence of the Eurocentric view on the historical development of mathematics. (Ibid., pp. 11–12)

George Joseph is criticizing the dominance of a Eurocentric (and male) mathematical hegemony which has created a judgmental situation within the discipline whereby, for example, deciding what constitutes powerful mathematics, or when a proof proves and what form a rigorous argument takes, is dictated and reinforced by those in influential positions. How often do we hear statements, often made about a geometric proof, dismissing it as 'merely a demonstration' or the suggestion that computer-assisted proofs are not quite as 'good' as those developed without a computer? How frequently are students encouraged to believe that the mathematico-scientific and technological development of the West has been made independently of a systematic knowledge and resource exploitation of the rest of the world? The colonization of mathematics has been so successful that the history of their own mathematical culture and its contribution to knowledge is often unknown to students in Africa, Asia and Latin America. Such bias is increasingly under attack (see, for example, Joseph, 1991; Needham, 1959; Nelson *et al.*, 1993; van Sertima, 1986; Zaslavsky, 1973) as researchers uncover the richness and power of mathematical and scientific development in the non-European world which has been obscured by the rewriting of history from a European perspective. If the body of knowledge known as mathematics can be shown to have been derived in a manner which excluded non-Europeans and their mathematical knowledge, why not conjecture that the perceived male-ness of mathematics is equally an artefact of its production and its producers?

Since I am arguing that mathematics is socio-cultural in nature, the conditions under which it is produced are factors in determining the products. 'Important' mathematical areas are identified, value is accorded to some results rather than others, decisions are taken on what should or should not be published in a society determined by power relationships, one of which is gender. Mathematical products can then be seen as the outcome of the influence of a particular 'reading' of events at a given time/place. Such readings are referred to by Sal Restivo (1992) as 'stories about commercial revolutions and mathematical activity, as in Japan, or about the "mathematics of survival" that is a universal feature of the ancient civilizations. And they can be stories about how conflict and social change shape and reflect mathematical developments' (p. 20). In a plenary lecture given at the 1994 American Educational Research Association Conference, Jerome Bruner pointed out that explanation as causal is a post-nineteenth century phenomenon. A longer history can be found for interpretation the objective of which is understanding and not explanation. He made out a case for understanding being viewed as both contextualizing and systematizing and he advocated a route to contextualizing in a disciplined way through narrative. From this perspective, codified mathematics can be viewed as reified narrative and it no longer seems so absurd to ask how different narratives, or stories, might constitute alternative mathematics (in the plural). Mathematics as a particular form of story about the world *feels*, to me, very different from mathematics as a powerful explanation or tool. Re-telling mathematics, both in terms of context and person-ness, would consequently demystify and therefore seem to offer opportunities for greater inclusivity.

Knowing Science and Mathematics

The feminist literature on the philosophy of science I find very valuable for the clarity with which it has sharpened the critical debate on the nature of knowledge in science and how that knowledge is derived. However, it is noticeable that the content criticisms of science are rooted in the empirical disciplines. For example, female primatologists such as Jane Goodall (1971), Dian Fossey (1983) and Sarah Hrdy (1986) challenged conceptions of interactive behaviour by refusing to accept the prevailing (male) views on dominance and hierarchy in sexual selection. Evelyn Fox Keller (1983) described Barbara McClintock's approach to her study of maize as highlighting a symbiotic relationship between the plant and its environment which was distinctively different to the more usual 'objective' investigation undertaken by botanists. Rachel Carson (1962) is frequently cited for her early work on ecology and the broad view that she took about the environmental effects of pesticides. In all these cases, the results of the science were different from what had, formerly, been expected because different questions were asked about what was being observed and different methods were used to make the observations.

However, criticisms of, for example, nuclear physics are more likely to focus upon the social effects of the science, rather than the science itself (see, for example, Easlea, 1983). This is not to diminish the importance of developing models of scientific use and abuse which criticize the purposes, products and implications of scientific developments. But, as with mathematics, it is difficult to confront the abstractions which are the substance and tools of the discipline and the methods used in their derivation especially where these are analytic and non-observational in order to ask what differences a female perspective would make to them.

In what ways might the questions, or the styles of enquiry or the mathematical products differ if mathematics were accepted as a socio-cultural construct? Part of the difficulty in responding to this question resides in the highly successful socialization experiences through which we all go in order to achieve success at mathematics. It is exceedingly difficult to dismantle the beliefs which have been integral to our learning experiences of mathematics and almost impossible to construct in our imaginations alternatives to the processes which we have been taught and with which we have gained 'success'. Hence, scratch a pedagogical or philosophical constructivist and underneath you are likely to expose an absolutist. In other words, it might be acceptable to negotiate a curriculum or introduce a collaborative, language-rich environment within which to make the learning of mathematics more accessible, but the mathematics itself is considered non-negotiable. However, to be consistent in our critiques, we cannot avoid addressing the nature of knowing mathematics along with the philosophy and pedagogy of the discipline.

Knowing mathematics, and science, has traditionally required entry into a community of knowers who accord the status of 'objective', 'in some sense eternal and independent of the flux of history and culture' (Restivo, 1992, p. 3), to the knowledge items as well as to the means by which these items are derived. However, 'objectivity is a variable; it is a function of the generality of social interests. Aesthetic and truth motives exist in the realm of ideas, but they are grounded in individual and social interests ranging from making one's way in the world (literally, surviving) to exercising control over natural and cultural environments' (Ibid., p. 135). A consequence of this is that 'a mathematical object . . . like a hammer or a screwdriver, is conceived, constructed, and put to use through a social process of collective representation and collective elaboration' (Ibid., p. 137). If we are to argue for a different conception of mathematical knowing from that traditionally accepted, we must address the meaning which is to be understood by 'objectivity' since the truth status accorded to mathematical objects underpins the pervading epistemology. Criticizing scientific 'objectivity' along similar lines, Sandra Harding (1991) has called for an epistemology of the sciences which requires a more robust standard for objectivity than that currently in use. This would include the critical examination, *within scientific research* (Ibid., author's italics, p. 146),

> of historical values and interests that may be so shared within the scientific community, so invested in by the very constitution of this or that field of study, that they will not show up as a cultural bias between experimenters or between research communities. (Ibid., 1991, p. 147)

And she further noted that:

> the difficulty of providing [such] an analysis in physics or chemistry [and, I would add, mathematics] does not signify that the question is an absurd one for knowledge-seeking in general, or that there are no reasonable answers for those sciences too. (Ibid., p. 157)

Sue Rosser, in her book, *Female-Friendly Science* (1990), used women's experience of knowing and doing science to draw out differences from what she called the conventional androcentric approaches. Amongst many of the possible inclusionary methods she listed in Chapter 5 are:

- expanding the kinds of observations beyond those traditionally carried out;
- increasing the numbers of observations and remaining longer in the observational stage of the scientific method;
- accepting the personal experience of women as a valid component of experimental observation;

- being more likely to undertake research which explores questions of social concern than those likely to have applications of direct benefit to the military;
- working within research areas formerly considered unworthy of investigation because of links to devalued areas;
- formulating hypotheses which focus on gender as an integral part; and
- defining investigations holistically.

This list, useful as it is for science, does not generalize easily to mathematics although the links to the history, philosophy and pedagogy of mathematics are more obvious. But help appears to be at hand.

Being a Mathematician

In *The Emperor's* [*sic*] *New Mind*, Roger Penrose (1990), arguing from the powerful position of a research mathematician at the top of his profession, claimed that the mathematician's 'consciousness' is a necessary ingredient to the comprehension of the mathematics. He said:

> We must 'see' the truth of a mathematical argument to be convinced of its validity. This 'seeing' is the very essence of consciousness. It must be present *whenever* we directly perceive mathematical truth. When we convince ourselves of the validity of Gödel's theorem we not only 'see' it, but by so doing we reveal the very non-algorithmic nature of the 'seeing' process itself. (Penrose, 1990, p. 541)

Elsewhere in his book, and in contradiction to the above, Roger Penrose supported a Platonic approach to mathematics in that he propounded a discovery, rather than an invented, perspective on the discipline. That is, mathematics is out there waiting to be uncovered rather than within the head (and possibly the heart?) of the mathematician. And yet, Roger Penrose himself admitted that 'seeing' the validity of a mathematical argument must be a personal experience and one which, it seems reasonable to me to assert, can be assumed to differ between individuals. By arguing that 'seeing' is non-algorithmic, Roger Penrose permitted the personalization of the process. He reinforced this with the statement:

> There seem to be many different ways in which different people think — and even in which different mathematicians think about their mathematics. (Ibid., p. 552)

However, for me, far from accepting that the outcomes of mathematical thinking are discovered mathematical 'truths', the inevitable conclusion of his statement is that there are potentially many different mathematics.

The contradiction would appear to lie in a different perspective on the mathematician than on mathematics itself. Roger Penrose viewed a mathematical statement, once articulated, as being absolute, that is either right or wrong, and its status verifiable by any interested party. But he said, in the

> conveying of mathematics, one is *not* simply communicating *facts*. For a string of (contingent) facts to be communicated from one person to another, it is necessary that the facts be carefully enunciated by the first, and that the second should take them in individually ... the *factual* content is small. Mathematical statements are necessary truths ... and even if the first mathematician's statement represents merely a groping for such a necessary truth, it will be that truth itself which gets conveyed to the second mathematician ... The second's mental images may differ in detail from those of the first, and their verbal description may differ, but the relevant mathematical idea will have passed between them. (Ibid., p. 553)

Despite the assumed personal nature of the communication and the expectation of differences in human images and descriptions, there is an assumption that the 'mathematics', the essential 'truth' of the statement, can and will be the same for all. This is repeatedly refuted by the message of many of the anecdotes which are recounted by, and about, mathematicians. For example, how is it possible to interpret the kind of intuitive insights which Roger Penrose himself, and other mathematicians such as Poincaré, Hadamard, Thom, claim to have had and which have led to their finding particular, personal resolutions of a mathematical problem. Given that it is reasonable to expect that any one problem might be amenable to a number of different routes for solution, an individual is likely to fall on the one which matches his or her experience, approach, preferences, possibly making the mathematical outcome different from that which would be offered by another individual. Of course, once articulated, the internal consistency of the mathematical argument is verifiable. The most recent attempt to prove Fermat's Last Theorem provided an example of the unverifiability, by most mathematicians, of the claims being made and, consequently, both the potential non-uniqueness and fragility of their status. And, even if the internal consistency is substantiated, this does not additionally encompass any objective status nor any implication of uniqueness, it seems to me. The social context within which the mathematics is placed does, however, offer one explanation for apparent uniqueness, or at least convergence of 'solutions', given that it describes and constrains the 'possible'. Thus, a piece of mathematics is both contributory to, and defined by, the context within which it is derived.

A belief in the world of mathematical concepts existing independently of those who develop or work with them is attached to embracing the 'objective' truths of mathematics. An image of 'variable' truth, that is degrees of correctness, or solutions responsive to different conditions, is unacceptable to many within the discipline despite the support from the history of mathematics that understandings change over time as the foci and the current state of knowledge change. The social context of a mathematical statement, the impact upon it of the interests, drives and needs of the person deriving and then communicating it, are dismissed by many mathematicians as inappropriate to the product. Thus, the distinction is made between the person who is working at the mathematics, and the mathematics itself. But I believe that Roger Penrose failed to sustain this distinction particularly in his discussion of intuition, insight and the aesthetic qualities of mathematical thinking. He underlined person-ness by reiterating an argument, (see, for example, Thom, 1973, pp. 202–6) that:

> the importance of aesthetic criteria applies not only to the instantaneous judgements of inspiration, but also to the much more frequent judgements that we make all the time in mathematical (or scientific) work. Rigorous argument is usually the *last* step! (Penrose, 1990, p. 545)

In drawing a close analogy between mathematical thought and intuition and inspiration in the arts, he added:

> The globality of inspirational thought is particularly remarkable in Mozart's quotation (from Hadamard, 1945) 'It does not come to me successively . . . but in its entirety' and also in Poincaré's 'I did not verify the idea; I should not have had time'. (Ibid., p. 347)

Any de-personalization of the mathematical process and reification of the product pushes mathematics back into the absolutist position by objectivizing the 'truths'. However, accepting a mathematics which is not absolute, is culturally defined and influenced by individual and social differences is not only of great interest to those who have argued for an inclusive mathematics but challenges the discipline epistemologically as well as philosophically and pedagogically.

> It does not seem untimely to suggest a theory of knowing that draws attention to the knower's responsibility for what the knower constructs. (von Glasersfeld, 1990, p. 28)

Once we refocus from knowing *that* a particular mathematical outcome exists to knowing *why* that outcome is likely under particular circumstances, we are distinguishing between the 'objective' knowledge of the outcome and the 'subjective' knowing which underlies how to achieve that outcome. This begins to be familiar as the old debate between product and process. However, by attempting to construct a theory of knowing, I am moving past the false dichotomy of product/process which polarized the how and the what, towards a reconceptualization and integration of the how with the what. The value to pedagogues of such an approach is obvious. As teachers, we can recognize when a learner mimics a piece of mathematical behaviour rather than acquires it as his or her own. The articulation of an epistemological position on knowing mathematics which is predicated on mathematical enquiry, rather than receptivity, challenges teacher behaviour. Rather than demanding evidence of the acquisition of mathematical objects by students, it assumes that mathematical behaviours and the changes in behaviour that might signify learning are products of, and responsive to, the community within which the learning is situated. Recounting different narratives, speculating about their similarities and differences, querying their derivations and applications, denies 'objectivity' and reinstates the person and the community in the mathematics. Such reconsideration of the characteristics of science and mathematics has underpinned much of the feminist work in the philosophy of science already referenced and is exemplified in the work of Suzanne Damarin. In an article (1991), she presented a table of generalized descriptors 'not as a definitive description of feminist science, but rather as defining a tentative framework for examining whether and how the teaching of science might be made more consistent with feminist conceptions of science' (Damarin, 1991, p. 112).

As stated above, the philosophical challenge, while not necessarily acceptable to a large number of mathematicians, has been well formulated. (Reference has already been made to the work of Bloor, Davis and Hersh, Harding, Lakatos, and Restivo.) Gadamer (1975) added his argument that:

> all human understanding is contextual, perspectival, prejudiced, that is hermeneutic [and] fundamentally challenges the conception of science as it has been articulated since the Enlightenment. (cited in Hekman, 1990, p. 107)

as did Elizabeth Fee (1981) to:

> attack the objectivity that is part of the 'mythology' of science . . . [and] . . . re-admit the human subject into the production of scientific knowledge. (also cited in Hekman, 1990, p. 130)

Much of the pedagogic challenge is focused on the dysfunctional nature of the continuum between an absolutist philosophy of mathematics and a transmissive pedagogy and the poverty of the product/process distinction:

> On the one hand, authors and publishers produce textbooks that do not have to be read before doing the exercises; on the other hand, teachers acquiesce by agreeing that this is the way mathematics ought to be taught . . . the real importance lies not in the students' ability to conceptualize, but rather in their ability to compute. Teachers tend to underscore this by their rapt attention to correctness, completeness, and procedure. Students comply with the grand scheme by establishing as their local goal the correct completion of a given assignment and as their global goal receiving their desired grade in the course. For most, once it's over, it's over. (Gopen and Smith, 1990, p. 5)

Compare this with a student-centred problem-solving approach:

> An instructor should promote and encourage the development for each individual within his/her class of a repertoire of powerful mathematical constructions for posing, constructing, exploring, solving and justifying mathematical problems and concepts and should seek to develop in students the capacity to reflect on and evaluate the quality of their constructions. (Confrey, 1990, p. 112)

Researchers have argued that creative mathematicians are more likely to develop by encountering and learning mathematics in a classroom climate which supports individuals within social groupings; that the negotiation of meaning both within the group and between the group and conventional social understandings needs to be encouraged (see for example Davis *et al.*, 1990). These philosophical and pedagogical critiques, in my view, would be strengthened by the focus, structure and consistency which is gained from an epistemological stance, that is, a formulation of the nature of knowing mathematics.

The Epistemological Challenge

I believe that we can discern the outline of an epistemological challenge to mathematics which, potentially, incorporates approaches consistent with, and familiar to, broader constituencies than European, middle-class males. These approaches are inclusive, rather than exclusive, accessible rather than mystifying, encompassing of as wide a range of styles of understanding and doing mathematics as possible rather than reducible to those styles currently validated by the powerful. I am claiming that knowing, in mathematics, cannot be differentiated from the knower even though the knowns ultimately become public property and subject to public interrogation within the mathematical community. Knowing, however:

> involves encouraging rebellious spirits to blossom with free rein to the imagination, preserving a certain nimbleness of mind while affording it the means of being creative. The 'training' procedures, as we conceive them and ordinarily practise them, hardly lend themselves, one must admit, to that kind of enticement, since they more often emphasize the transmission of acquired knowledge and apprenticeship in proven methods. And considering that those procedures resemble an obstacle course where the competition is tighter and tighter, this hardly encourages departing from the beaten path. (Flato, 1992, p. 75)

I am speculating that five categories, drawn from the work already cited and consistent with the above critique, might distinguish the ways in which (creative) mathematicians come to know mathematics and that, in their choice of mathematical areas to pursue, more women (and men) might feel comfortable with an epistemology of mathematics described in this way. The assumption is that such an epistemology would displace dualisms such as the relativist/absolutist dichotomy and expectations of a value-free mathematics with an hermeneutic and pluralist approach. It would open the way towards an inclusive perspective on mathematics by challenging our understanding of what constitutes mathematics.

I propose defining knowing in mathematics in relation to the following five categories derived from the reading reviewed above in the philosophical, pedagogical and feminist literature:

- its person- and cultural/social-relatedness;
- the aesthetics of mathematical thinking it invokes;
- its nurturing of intuition and insight;

- its recognition and celebration of different approaches particularly in styles of thinking; and
- the globality of its applications.

Knowing mathematics would, under this definition, be a function of who is claiming to know, related to which community, how that knowing is presented, what explanations are given for how that knowing was achieved, and the connections demonstrated between it and other knowings (applications). What evidence we have, usually sited in the learning and assessing of school mathematics, suggests that inviting students to define and describe their knowing in mathematics in these ways does have gender implications (see, for example, Burton,1990b; Forgasz, 1994; Stobart *et al.*, 1992).

The similarities with Sue Rosser's (1990) and Suzanne Damarin's (1991) lists of the differences between male- and female-friendly science are encouraging. For example, both refer to the expansion of the kinds of observations carried out, the recognition of, and concern for, personal responsibility and the consequences of actions. I have listed a valuing of intuition and insight and the recognition and celebration of different approaches. Globality, or in both Sue Rosser's and Suzanne Damarin's terms 'holism', is a feature. A need to accept the personal experience of women as a valid component of experimental observations is acknowledged where I have pointed to person-relatedness which is important to knowing mathematics. Susan Hekman's (1990) analysis of the relationship between gender and post-modernism was also supportive of this approach both in drawing out the similarities in argument between feminists and post-modernists as well as pointing out the pervading influence of absolutism in affecting these stances. In Hilary Rose's words:

> A feminist epistemology ... transcends dichotomies, insists on the scientific validity of the subjective, on the need to unite cognitive and affective domains; it emphasises holism, harmony, and complexity rather than reductionism, domination and linearity. (Rose, 1986, p. 72)

The next step is to open a dialogue with practising mathematicians with a view to discussing the appropriateness of my description to their understanding of the nature of knowing in mathematics. This would be done in a style which would be rich in ethnographic data, encouraging the expression of feelings, aesthetics, intuitions, and insights. It would also attempt to challenge the effects of socialization into the mathematical culture in order to untangle differences from cultural similarities. Outcomes which are supportive of the suggested epistemological framework especially where these emphasize impact on gender inclusivity would provide a strong argument in favour of a re-perception and re-presentation of mathematics. The resulting narrative would have an internal consistency which should please all mathematicians.

Such anecdotal approaches as have already been made confirm the validity of the five categories in describing how mathematicians come to know. Arguments in favour of humanizing and demystifying the mathematics curriculum in schools have long been made with an implication that such attempts change perceptions of mathematics, and subsequent performance, by formerly underrepresented groups. However, these suggestions are rarely connected to epistemological frameworks of the discipline more frequently relating either to constructivist philosophy or empowering pedagogy. And Suzanne Damarin criticizes curriculum reformers for their 'reliance on the models of expertise and information processing, which are popular in current research on the cognitive bases of teaching and learning of science and mathematics [and] appears to be diametrically opposed to first-order implications of feminist pedagogical research' (Damarin, 1991, p. 108).

If the nature of knowing mathematics were to be confirmed as matching the description given in this chapter, the scientism and technocentrism which dominate much thinking in and about mathematics, and constrain many mathematics classrooms, would no longer be sustainable. Mathematics could then be re-perceived as humane, responsive, negotiable and creative. One expected product of such a change would be in the constituency of learners who were attracted to study mathematics but I would also expect changes in the perception of what *is* mathematics and of how mathematics is studied and learned. That such a possibility, in schools, is not outside the realms of possibility is suggested in Boaler (1993). We can also learn from experiences in other disciplines. English, for example, attracts predominantly female constituencies of learners at the undergraduate level, many of whom have been successful in developing academic careers. 'English was constructed as a liberal humanist discipline which demanded personal and thoughtful response . . . The most important characteristic of English, in the view of students and staff, is its individualism: the possibility of holding different views from other people' (Thomas, 1990, p. 173).

Providing a new epistemological context would enable the questioning of what mathematics is taught, how it is learned and assessed within a consistent treatment. 'By adopting an epistemological view of mathematical knowledge that stresses change, development, and its social foundations generally, and by consciously relating this to the curriculum process, the result would be to make the subject more open in its nature and more easily accessible' (Nickson, 1992, p. 131).

My aim in attempting this work is to question the nature of the discipline in such a way that the result of such questioning is to open mathematics to the experience and the influence of members of as many different communities as possible, thereby, I hope, not only enriching the individuals but also the discipline.

Note

1 This is a version of a paper first given at the IOWME study group at ICME-7 in Québec, 1992. Its present content owes much to discussion with, and comments from, members of IOWME. In addition, I would like to thank Mary Barnes, Leonie Daws, Stephen Lerman and the anonymous reviewers of ESM for challenging and provoking

reworking of the ideas. This paper is reprinted by permission of Kluwer Academic Publishers from *Educational Studies in Mathematics*, 28, 3, Special Issue on Gender, edited by Gilah Leder, to appear in Spring 1995.

References

BLOOR, D. (1991) *Knowledge and Social Imagery*, 2nd ed., London, University of Chicago Press.

BOALER, J. (1993) 'Encouraging the transfer of "school" mathematics to the "real world" through the integration of process and content, context and culture', *Educational Studies in Mathematics*, 25, 4, pp. 341–73.

BRANNIGAN, A. (1981) *The Social Basis of Scientific Discoveries*, Cambridge, Cambridge University Press.

BRUNER, J. (1994) 'The Humanly and Interpretively Possible', Plenary address given to the AERA Annual Meeting, New Orleans.

BURTON, L. (Ed) (1986) *Girls Into Maths Can Go*, London, Holt Educational.

BURTON, L. (Ed) (1990a) *Gender and Mathematics: An International Perspective*, London, Cassell.

BURTON, L. (1990b) 'Passing through the mathematical critical filter — implications for students, courses and institutions', *Journal of Access Studies*, 5, 1, Spring, pp. 5–17.

CARSON, R. (1962) *Silent Spring*, New York, Fawcett Press.

CONFREY, J. (1990) 'What constructivism implies for teaching', in DAVIS, R.B., MAHER, C.A. and NODDINGS, N. 'Constructivist Views on the Teaching and Learning of Mathematics', *Journal for Research in Mathematics Education*, Monograph No. 4, Reston, VA, NCTM, pp. 107–22.

DAMARIN, S. (1991) 'Rethinking science and mathematics curriculum and instruction: Feminist perspectives in the computer era', *Journal of Education*, 173, 1, pp. 107–23.

DAVIS, P. and HERSH, R. (1983) *The Mathematical Experience*, Harmondsworth, Penguin.

DAVIS, R.B., MAHER, C.A. and NODDINGS, N. (Eds) (1990) 'Constructivist views on the teaching and learning of mathematics', *Journal for Research in Mathematics Education*, Monograph No. 4, Reston, VA, NCTM.

DEPARTMENT OF EDUCATION AND SCIENCE (1989) *Mathematics in the National Curriculum*, London, Her Majesty's Stationery Office.

EASLEA, B. (1983) *Fathering the Unthinkable: Masculinity, Scientists and the Nuclear Arms Race*, London, Pluto Press.

ERNEST, P. (1991) *The Philosophy of Mathematics Education*, Basingstoke, Falmer Press.

FEE, E. (1981) 'A feminist critique of scientific objectivity', *Science for the People*, 14, 30, p. 3.

FENNEMA, E. and LEDER, G. (Eds) (1990) *Mathematics and Gender*, New York, Teachers College Press.

FLATO, M. (1992) *The Power of Mathematics*, London, McGraw-Hill.

FORGASZ, H. (1994) *Society and Gender Equity in Mathematics Education*, Geelong, VA, Deakin University Press.

FOSSEY, D. (1983) *Gorillas in the Mist*, Boston, MA, Houghton Mifflin.

GADAMER, H.-G. (1975) *Truth and Method*, New York, Continuum.

GOODALL, J. (1971) *In the Shadow of Man*, Boston, MA, Houghton Mifflin.

GOPEN, G.D. and SMITH, D.A. (1990) 'What's an assignment like you doing in a course like this? Writing to learn mathematics', *The College Mathematics Journal*, 21, 1, pp. 2–19.

Hadamard, J. (1945) *The Psychology of Invention in the Mathematical Field*, Princeton, NJ, University Press.
Harding, S. (1986) *The Science Question in Feminism*, Milton Keynes, Open University Press.
Harding, S. (1991) *Whose Science? Whose Knowledge?*, Milton Keynes, Open University Press.
Hekman, S. (1990) *Gender and Knowledge — Elements of a Postmodern Feminism*, Boston, MA, Northeastern University Press.
Hirst, P.H. (1965) 'Liberal education and the nature of knowledge', in Archambault, R.D. (Ed) *Philosophical Analysis and Education*, London, Routledge and Kegan Paul.
Hirst, P.H. (1974) *Knowledge and the Curriculum*, London, Routledge and Kegan Paul, pp. 113–38.
Hrdy, S.B. (1986) 'Empathy, polyandry, and the myth of the coy female', in Bleier, R. (Ed) *Feminist Approaches to Science*, Oxford, Pergamon Press, pp. 119–46.
Hull, R. (1985) *The Language Gap*, London, Methuen.
Joseph, G.G. (1991) *The Crest of the Peacock*, London, Tauris and Co.
Keller, E.F. (1983) *A Feeling for the Organism: The Life and Work of Barbara McClintock*, New York, W.H. Freeman.
Keller, E.F. (1985) *Reflections on Gender and Science*, New Haven, CT, Yale University Press.
Kelly, A.V. (1986) *Knowledge and Curriculum Planning*, London, Harper and Row.
Kelly, A. (Ed) (1987) *Science for Girls?*, Milton Keynes, Open University Press.
Lakatos, I. (1976) *Proofs and Refutations*, Cambridge, Cambridge University Press.
Lakatos, I. (1983) *Mathematics, Science and Epistemology*, Cambridge, Cambridge University Press.
Leder, G. and Sampson, S.N. (Eds) (1989) *Educating Girls*, Sydney, Allen and Unwin.
Needham, J. (1959) *Science and Civilization in China*, Cambridge, Cambridge University Press.
Nelson, D., Joseph, G.G. and Williams, J. (1993) *Multicultural Mathematics*, Oxford, Oxford University Press.
Nickson, M. (1992) 'Towards a multi-cultural mathematics curriculum', in Nickson, M. and Lerman, S. (Eds) *The Social Context of Mathematics Education: Theory and Practice*, London, South Bank Press, pp. 128–33.
Penrose, R. (1990) *The Emperor's New Mind*, London, Vintage.
Polanyi, M. and Prosch, H. (1975) *Meaning*, London, University of Chicago Press.
Restivo, S. (1992) *Mathematics in Society and History*, Dordrecht, Kluwer.
Rose, H. (1986) 'Beyond masculinist realities: A feminist epistemology for the sciences', in Bleier, R. (Ed) *Feminist Approaches to Science*, Oxford, Pergamon Press.
Rose, H. and Rose, S. (1980) 'The myth of the neutrality of science', in Arditti, R., Brennan, P. and Cavrak, S. (Eds) *Science and Liberation*, Boston, MA, South End Press.
Rosser, S.V. (1990) *Female-Friendly Science*, New York, Pergamon.
Stobart, G., Elwood, J. and Quinlan, M. (1992) 'Gender bias in examinations: How equal are the opportunities?', *British Educational Research Journal*, 18, 3, pp. 261–76.
Thom, R. (1973) 'Modern mathematics: Does it exist?' in Howson, A.G. (Ed) *Developments in Mathematics Education*, Cambridge, Cambridge University Press, pp. 194–209.
Thomas, K. (1990) *Gender and Subject in Higher Education*, Society for Research in Higher Education and Open University Press, Buckingham.

van Sertima, I. (Ed) (1986) *Blacks in Science: Ancient and Modern*, New Brunswick, NJ, Transaction.

von Glasersfeld, E. (1990) 'An exposition of constructivism: Why some like it radical', in Davis, R.B., Maher, C.A. and Noddings, N. (Eds) 'Constructivist Views on the Teaching and Learning of Mathematics', *Journal for Research in Mathematics Education*, Monograph No. 4, Reston, VA, NCTM, pp. 19–29.

Whyte, J., Deem, R., Kant, L. and Cruickshank, M. (Eds) (1985) *Girl Friendly Schooling*, London, Methuen.

Zaslavsky, C. (1973) *Africa Counts: Number and Pattern in African Culture*, Boston, Prindle, Weber and Schmidt.

Chapter 25

Mathematics: An Abstracted Discourse

Betty Johnston

It is clear that we wouldn't have written this book and you wouldn't be reading it if we weren't all concerned in some way about 'the gender imbalance in mathematics'. What is not so clear is just how we understand this imbalance, what we believe its causes are, and what we think we should do about it.

I have questions about this imbalance, which I cannot address here. How does the imbalance manifest itself? What 'facts' are we using to help us see it, who collected them, for what purpose, on what evidence? What does 'good at maths' mean and how do we measure it? How do we construct our understanding of the 'facts'? How do we use it? And, finally, why do we care so very much that everyone should do mathematics? These are all questions addressed elsewhere by people such as Yves Chevallard (1989), Richard Noss (1991), Dorothy Smith (1990), and Valerie Walkerdine (1988).

All sorts of reasons have been proposed to account for the indisputable, widespread fear and dislike of mathematics. Some of these reasons are specifically related to women, others are more generally applicable to groups in our society who miss out — all are variations, more or less sophisticated, on the theme 'you must have been away at some point', or 'you must have had a bad teacher', or 'perhaps you thought it was unfeminine', or 'maybe you haven't got a mathematical mind'. And while all this weaves into the complex pattern that is our experience, perhaps we don't take seriously enough the voices that say, again and again, 'but it doesn't make sense', and 'what's the point of it?' Perhaps what they are saying simply *is* true. Perhaps mathematics, their mathematics, secondary-school mathematics, doesn't make sense. Perhaps the fault is in the mathematics, and not the teaching, not the learning, not the people. At the very least, it is a question worth focusing on for a while.

Mathematics is a human construction and learning mathematics is a social process (Ernest, 1991). Western society is saturated with numbers. Quantification, rationality and abstraction are intricately connected, increasingly embedded and highly valued in the scientific and technological world. What I am interested in is the interaction between mathematics and society, in how, for instance, the quantification of society constructs us particularly, as women, and in how we construct or resist it. As Sue Willis (1989) argues, attempts to explain the gender imbalance shift the focus from *girls can't* (they are biologically incapable of doing so), to *they don't* (they are not socialized into that particular role), to *they won't* (they choose not to). The emphasis changes first from biological determinism to a more complex social determinism, and, from this still quite passive picture of women, to one that insists that individuals are active participants in their formation as mathematical beings.

Two theorists who have grappled with the question of agency within structure are the Canadian sociologist, Dorothy Smith, and the German sociologist and psychologist, Frigga Haug. Both work from within feminist and socialist frameworks; both begin from, and value, the standpoint of women; both develop methodologies that allow examination of the interface between the everyday world and the wider social structures. Both approaches would be valuable tools for our research, but I will focus here on the use of Haug's memory-work.

In examining the process of female sexualization, Haug (1986) developed a methodology that has become known as memory-work. In the last five years, variations of memory-work have become quite widely used in a number of research projects worldwide. In Haug's original use of the method, a group of people interested in a given topic, and preferably from a variety of backgrounds, jobs, and disciplines, meet regularly. They write and analyse relevant 'memories' from their own lives; they collect and examine other materials — documents both old and new, dogmas, fairy tales, proverbs, newspaper articles; they examine and discuss theories and opinions about the process they are focusing on. They do all this from a theoretical viewpoint that seeks to use experience as the basis of knowledge and refuses to view people simply as bearers of roles, but sees them as active participants even perhaps in their own subordination. Gender, like class and ethnicity, is neither chosen nor given, but produced and therefore transformable. Memory-work differs from autobiography, being not an investigation of the way things *really* were with continuities retrospectively constructed, but a study of the processes through which we have become the people we are today. Haug and others recognize that individuals do not give objective accounts of themselves. But, they say, our reinterpreting, falsifying, forgetting and repressing do not need to be seen as obstacles to the truth, rather they can become opportunities for investigation. In examining how individuals construct their identity, and what becomes subjectively significant, we are examining how women grow into the structures of society (Haug, 1986, p. 40). The emphasis of the investigation is not on controlling others, but on taking control for ourselves.

I also wanted to examine the action of people within given social structures, to try to 'identify points at which change is possible, where a shift from heteronomy (subjection to external laws) to autonomy can take place' (Haug, 1986, p. 41). To do this I have used the process of memory-work with a group of women in Sydney to reconstruct the process of mathematization through an examination of our own past experiences. We tried to tease out the ways in which the discourse of mathematics, all the varied knowledge, practices and beliefs surrounding mathematics, works to insert us into its world, whether as successes or not, and, how we, as individuals, weave ourselves into the mathematical world. We met for two or three hours every two weeks, usually on a Sunday morning, over a late breakfast, for a period of four months. The women involved in the group were, as in Haug's original group, from a variety of educational and work backgrounds — a high-school science teacher, an adult-literacy teacher, a librarian, an engineer, a design student, a numeracy lecturer, a research assistant. This variety brings a range of experience and theoretical frameworks to the discussions and analysis, but we were not randomly chosen and so in no sense could we be seen as 'representative' of our particular backgrounds. Like Haug (1986), we are not using memory-work to make statistically valid statements, so much as to explore possible relations and dynamics, and possible transformations.

The memories and our discussions have revealed a wilderness of starting points: an absence of mothers and a persistence of fathers; a felt connection between sin and correctness; a surprising sensuousness associated with many early memories of doing mathematics; widely differing understandings of the meaning of 'cleverness'; mathematics as order, and order shifting between pattern, sequence and command; the violence of measurement; measurement and ac*count*ability. . . . In this chapter I trace one strand that makes connections between gender imbalance and the discipline of mathematics.

Listen to three of the stories written in response to the cue: 'doing maths'. Members of the group were asked to write about an early experience of doing mathematics, and to write it in the third person in an effort to distance themselves and to prevent them censoring the stories, leaving out what might be significant connections.

Kate
Her memories are more impressions: maths was part of the fluid world of being in a cocoon which was school — impressions of warmth, laughter, friends, the certainty of the sums and the rules. Chaos was not a part of this world. Order, cut grass, humming summer heat and security.

Alison
8.30 am. Convent school under a tree. The three girls are animatedly discussing their maths homework from the previous night. Not set homework. This was stuff they'd decided to do, they couldn't wait to do, a well-worn mathematics pathway that had been trodden by thousands before them, a journey of discovery . . . The usual hum of maths conversations for these three students. Lynne could not get out problem 56, a tortuous piece of geometry which had left a smudgy trail of carbon pencil over the page. The same page was already indented by the pressure of earlier problems, so artfully deciphered that their presence could be felt as your hand progressed across the page. What a turn on! Sheila had fathomed 56, and proudly unveiled her solution . . . Lynne can't believe she didn't think up the solution herself — shit! With a flurry of rubbers, Lynne's page is erased. The now famous solution is recorded. The page is now even more smudged and creased — a beautiful record of human endurance and application. It looks fantastic! This

is the best part — what visuals! Their own shared mathematical aestheticism.

Marie
She's about 11-years old, crouched over her work, arms around the outskirts of her books, hiding it. Palms sweating. The extent of her shame, not knowing, dirtiness, must somehow be contained and hidden. Then there is a figure behind her, towering, menacing: Sister Peter. Her voice is cold and low and in a few words the shame, not knowing, dirtiness of the young girl are exposed, public. 'That's wrong. What do you think you're doing. Tear it out and start again.' Repeat the shame, repeat the humiliation.

What links were being made? What silences were apparent? One link that surprised us was the strong sensuousness of the stories, from Kate's ease and summer, to Alison's crumpled paper and messy pencil which she later described as the physical representation of thinking, to Marie's connection of failure with mess, dirt and towering authorities. A silence that we noticed, after some time, was that none of the stories actually contained any mathematics. It seemed that not only had we learnt *in* context, we had learnt *a* context: it was the whole context that was remembered, though, except for Alison who had retained a little, the mathematics had fallen out. Why was the mathematics itself lost?

An Ordered Daily Training . . .

The point I want to follow here is our realization that the stories seemed to be not descriptions of a single occasion so much as a condensation of a number of repeated similar experiences, layered memories. Marie's memory, for instance, reflected the continuing, everyday pettiness and horror of her first year at high-school; what we are seeing in each story is an ordered daily training towards a particular mathematical normality.

. . . By the Reduction to Mathematics of the Everyday World

Much of the mathematics that most people use routinely is concerned with measurement, with the breaking up of wholes into parts. There is hardly an aspect of our lives free from this quantification. The history of the measurement of time, for instance, demonstrates its increasing use as a means of controlling working populations (some schools in New South Wales have teaching periods of 52.5 minutes!). Are there differences in how men and women experience this quantification? Why, as Alison asked, did boys' watches have a second hand, while girls' only had hours and minutes? One strand in the struggle to be an active participant in her own life emerges for Marie from the unravelling of her memories about time, and connections between paid work and the measurement of time:

> Part of the struggle for me around work has been that I feel like I never do a good enough job and I'd be much happier working for nothing, because I can be myself, and in fact my experiences with teaching absolutely confirm that ... that when late last year, I worked with somebody for nothing, well for an exchange of energy — she's going to do work on my garden — it was great, but when I work for money in my job full-time I ... feel totally distressed, all the time, ... and a feeling that I have a lot at work is that my time isn't my own ... there's all this funding pressure now to take particular kind of people into our groups who will be able to articulate into the next exciting episode of TAFE, in six months time ... it has been fairly open-ended, it depends on people's goals, but now it's well a thing of people can't spend as much time ... but for a lot of people, yeah, that's what they are wanting, that sort of control over their own learning and ... confirmation and stuff, but it actually takes time to do that.

Women's work, paid and unpaid, is often described as a 'double burden'. Frigga Haug (1992, p. 260) argues that it is more accurate to say that 'women are located in two areas with contradictory logics of time', the measured time of paid work and the unpaid, and therefore unvalued, situations where spending more time is better than rationalization. And so, she suggests, a certain resistance to mathematical thinking may be part of women's upbringing in order to prevent schizophrenia.

... By the Separation of Mathematics from the Everyday World

For Kate also, the normality constructed in the process of mathematization, involved a not-belonging, a separation of thought and experience.

> So maths geniuses are young because ... because maths is so ... thin, and they can say: I can understand it, because they think they only need to understand that much. I mean they are just totally unaware of what they don't understand ... there are people who are very good at spotting patterns, and then there are other people, but they are very rare, who when they look at a page they can see a pattern and it actually has a meaning in terms of reality for them ... the only way I can describe it is it has a 3-D reality, it's a real thing that's happening, but they can also see a pattern ... and I personally think the people who are not good pattern spotters are people who right from an early age have demanded that level of understanding, rather than it being the other way round ... like I've been able to do maths because I've accepted that I shouldn't ask questions ... I knew that I had pattern-spotting skills but suddenly they weren't working and so why weren't they working and it linked into this whole thing about not having this history of 3-D objects and in practical

> situations linking with patterns ... I mean a lot of the guys haven't made the connections between their experiences and pattern making but they do during the course ... suddenly the course is making sense of their reality. ...

Like Kate, none of the women in the group had had the experience of these men in the engineering course, of mathematics reinforcing their reality.

... In the Normality of Heteronomy

As Marie's early stories unravelled, at times very painfully, we often recognized ourselves. 'With maths, everything had to be laid out and ordered and mine just never was ... and also you knew there was a right answer ... that I couldn't necessarily arrive at through my own process ... the right and wrong dichotomy is so strong and then ... there's a link with sin.'

Her sense of the external structures defining her responses, of herself as a victim, continued into high-school.

> I see school as a quite profound violation ... I think I'm a very skilled mimic and what started out as a way of surviving, picking up what was socially accepted behaviour and just reproducing it ... while leaving something intact inside ... I think I actually really lost myself ... My memories of maths at school, was that it was a construction that other people made for me and of me, but that I never shared myself with ... and because I feel like I was mimicking and that things were not coming from within, then I have always had a very deep feeling of being found out one day ... that one day the axe would fall down ... and I actually just remembered today that I did all right in mathematics, like I got an A for School Certificate [fourth year of high-school], and thinking, yes I got away with that, but I might not be able to sustain the act.

To learn every day that it is normal that mathematical knowledge is externally given and monitored, that patterns reflect no reality, that a quest for a certain kind of understanding hinders success, that everyday practices are quantified and regulated by a vast array of indices, is to experience mathematics as a profoundly decontextualized discourse: an abstracted discourse that could refer to anything, and for most people refers to nothing (Walkerdine, 1992). It is to experience mathematization as the ordered daily training in the normality of heteronomy.

The abstract mathematics that most of us know, like economics, ignores both its sources and its applications and consequences. That industrialized society should have such decontextualized mathematics is no accident, Peter Denny (1986), in a study that compares the mathematics of a hunter society, the Ojibway and Inuit peoples of North America, with that of agricultural and industrial societies, has argued:

> [Our content-free] style of mathematics mirrors the style that is required

> by industrial technology. . . . the isolation of crucial variables and their manipulation independent of context is needed for the high degree of alteration of the environment achieved by industrial society . . . It is inappropriate however for people in hunting societies who earn their living from a relatively unaltered environment . . . Since [the hunter] will not be controlling nature, isolated knowledge of crucial variables will not help, but inclusive knowledge of the whole pattern of natural processes will be imperative. (Denny, 1986, p. 142)

The difference between the two kinds of mathematics is clear in the case of navigation. The navigator in industrial society can derive a direction from a single selected factor, magnetic north, and maintain this information in isolation by using a gyroscopic compass even if the magnetic field becomes disturbed. The navigator in a hunting society must pay attention to dozens of factors simultaneously. The Inuit hunter travelling by dog-sled through a blizzard must register changing conditions of snow, ice, wind, temperature, and humidity, all in large complex patterns, which, taken as whole structures, will successfully specify location. This high degree of contextualization can be achieved only if there is a general habit of treating information in context rather than in isolation. Therefore it will not surprise us if this inclusive style of thought shows up in the mathematics of hunting societies, just as the isolating style shows up in the mathematics of industrial society.

In the industrialized world, mathematics has become so 'content-free' that, for example, the metric system gives as standard measures of length, millimetres, centimetres and metres — the first two, too small, and the last, too large for comfortable measurement of many everyday objects. Feet and inches may have been less elegant, or less useful to science, but a few inches or a few feet effectively measured most things that ordinary people needed.

In modern, industrialized societies, with their proliferating global problems, increasingly complex and interrelated, such isolation is extremely dangerous and irresponsible. Abstract reasoning, Valerie Walkerdine (1992) argues, is not the ultimate pinnacle of intellectual achievement, but a massive forgetting which is intricately bound up with questions of power. Perhaps we should bear in mind Hegel's warning:

> [the analytical approach] labours under a delusion if it supposes that while analyzing the objects, it leaves them as they were: it really transforms the concrete into an abstract. And as a consequence of this change the living thing is killed: life can only exist in the concrete and one. Not that we can do without this division, if it is our intention to comprehend . . . The error lies in forgetting that this is only one half of the process, and that the main point is the reunion of what has been parted. (Hegel, 1975, p. 63)

From Heteronomy to Autonomy

If by 'discipline' we mean 'a branch of instruction or learning', then when we blame the discipline of mathematics itself for the gender imbalance, we are guilty yet again, in our essentially mathematical habit, of isolating a single variable and treating it out of context. We are guilty of inappropriate abstraction. To return it to its context we need to consider not only its knowledge, but its practices, its construction of meanings and emotions, and its constitution as a process that inserts us into given social worlds.

A step towards autonomy would be an appreciation of mathematics, not as the manipulation of the already abstracted, free-floating products of abstraction, but as the process of abstracting from the material context, and not only that, but of reconnecting to the concrete, to the context, as we are learning to do in ecology, learning to embed our knowing in an understanding of sources and consequences.

It was not until recently that Marie took a course about adult learning:

> . . . that's the only course that I've done that has in any sense come from me and touched me and it marked quite a profound personal shift . . . my first experience of that, of knowing in general and quite specifically in maths was with the maths anxiety workshop and the handshake problem . . . and this whole revelation, that it has a basis in concrete experience, that is if I understand the reality of it then I don't need to know the rules because I can just get there again . . . I could get back to the formula.
>
> And if you have people coming from their own experience, you have people who have self-confidence and are not fooled by Hewson and the rest of the politicians . . . and that's something I experienced in this course . . . this extraordinary sense of being in a process, that everything has meaning . . . and could be, like we're doing here . . . really enjoying using our own experience, no not using it, talking about it, valuing it to work on it, bringing together some ideas, to develop some knowledge . . . it is really exciting and sort of passionate and enjoyable, but we have people who are not socially controlled I think, in the end . . . and who come from being very damaged to increasingly feeling, I am worthwhile.

And so we travelled with Marie from a position as victim, from her training in the normality of heteronomy, to the possibility of ourselves also as actors, to the possibility of autonomy.

And What Are the Costs?

Even more fundamentally, maybe we should be asking a different question, not why is there an imbalance, why don't girls do as well as boys, but why does anyone ever do well in this 'thin' subject (as Kate called it)? And what are the costs, individually and socially, of the ordered daily training in the normality of heteronomy?

References

CHEVALLARD, Y. (1989) 'Implicit mathematics: Its impact on societal needs and demands', in MALONE, J., BURKHARDT, H. and KEITEL, C. (Eds) *The Mathematics Curriculum: Towards the Year 2000*, Perth, Science and Mathematics Education Centre, Curtin University, pp. 49–57.

DENNY, J.P. (1986) 'Cultural ecology of mathematics: Ojibway and Inuit hunters', in CLOSS, M.P. (Ed) *Native American Mathematics*, Austin, University of Texas Press, pp. 129–80.

ERNEST, P. (1991) *The Philosophy of Mathematics Education*, London, Falmer Press.

HAUG, F. (Ed) (1986) *Female Sexualization: A Collective Work of Memory*, translated into English by Erica Carter, London, Verso.

HAUG, F. (1992) *Beyond Female Masochism: Memory-Work and Politics*, London, Verso.

HEGEL, G.W.F. (1975) *Logic (Encyclopaedia of the Philosophical Sciences, Part 1)*, translated by W. Wallace (3rd ed.), Oxford, Clarendon Press, Section 38z.

NOSS, R. (1991) 'The social shaping of computing in mathematics education', in PIMM, D. and LOVE, E. (Eds) *The Teaching and Learning of School Mathematics*, London, Hodder and Stoughton, pp. 205–20.

SMITH, D. (1990) *The Conceptual Practices of Power: A Feminist Sociology of Knowledge*, Boston, MA, Northeastern University Press.

WALKERDINE, V. (1988) *The Mastery of Reason: Cognitive Development and the Production of Rationality*, London, Routledge.

WALKERDINE, V. (1992) 'Reasoning in a post-modern age', Paper presented at the 5th International Conference on Thinking, Townsville.

WILLIS, S. (1989) *'Real Girls Don't do Maths': Gender and the Construction of Privilege*, Geelong, Deakin University Press.

Chapter 26

Constraints on Girls' Actions in Mathematics Education

Marjolijn Witte

Introduction

The Dutch education system is based on a view that sees education as the transmission of knowledge, rather than as supporting the construction of integrated personalities, starting from individual differences. This emphasis on knowledge transmission creates a mathematics education that encourages students to model themselves after the experts who created mathematics. In this chapter, I argue that this is a view that also inhibits the participation of girls in mathematics.

Many theories have been advanced to explain why more girls than boys in western society drop out of mathematics education (Sherman, 1981; Pedro *et al.*, 1981; Leder, 1985). And, in spite of various attempts to change this situation, this pattern still persists (see, for instance, Sangster, 1988; Burton, 1990). Of course, the mere fact that girls prefer careers that avoid mathematics need not constitute a problem in itself — provided it is a choice. Given this, the question is, 'On what level, and what kinds of, intervention will be most effective in remedying the situation?'

Three Intervention Strategies

In this section, three different change strategies are distinguished. Most interventions fall into the first category which emphasizes individual choice and factors that determine this choice. The second category focuses on the educational environment, while the third conceives of mathematics education as a subsystem of society, influenced by a dominant language. This language puts constraints on the actions of learners and, as I will argue, disadvantages girls.

Improving the Student

The most frequently employed intervention strategy operates on the level of the individual girl. This strategy is based on hypotheses about what the average girl wants or can do. It may emphasize biological differences, such as girls having lower spatial-visualization abilities than boys, or the social construction of femininity as ways of behaving, choosing and acting. The aim of this strategy is to look for factors that might determine high gender-related drop-out rates and negative attitudes towards mathematics education, and use these factors to inform the design and implementation of interventions.

An example of this approach in the Netherlands is the governmental campaign, 'Choose Exact', which tries to influence girls' choices towards the mathematical sciences. According to many theories about girls' attitudes, girls lack a proper awareness of the usefulness of mathematics. This diagnosis has been translated into the notion that girls have to be convinced that well-paid jobs are more accessible if they choose to study the exact sciences. The campaign failed to reach its target. Certainly, girls became more aware of the relevance of the exact sciences, but this awareness did not change their choices (Ministerie van Onderwijs en Wetenschappen, 1990). Probably usefulness does not simply translate into career possibilities.

A more essential criticism of this approach is that it is too ambitious. The factors that seem to influence girls' choices are so many and so varied, and so interrelated, that choosing any collection of them on which to base an effective intervention is rather like buying a ticket in a lottery. A second drawback with this approach is that it focuses only on the girls and what shapes their behaviour in general. Very little attention is paid to their contact with mathematics as a discipline, as mediated by the educational system (for a feminist critique of the male character of much of science and mathematics see, for example, Keller, 1985; Harding, 1986; Walkerdine, 1989).

Improving the Environment

The basic idea behind the strategies in this category is to improve the educational environment in order to attract girls. This strategy, as applied in Dutch secondary mathematics education, has produced many changes in the curriculum and in methods of teaching. The most recent change was aimed at developing a form of mathematics education which was more open to all learners, and supporting them to the maximum of their abilities and interests. Any remaining differences, it was argued, would therefore be due solely to invariant differences at the individual level. This development, called 'realistic mathematics', deemed it important to build upon what pupils already know, their real-life experiences, the various concepts and procedures they have already developed. Hence, many different individual conceptions of 'mathematics' were assumed possible. However, in the initial trials, 'realistic mathematics' was not as successful as hoped. Boys were more attracted by the contexts chosen than were the girls whose interests and motivations were not encouraged as much as had been hoped (Dekker *et al.*, 1985; Vos, 1988). These so-called 'realistic contexts' seem not to appeal to girls' reality, and any attempt to make them more attractive to girls cannot succeed as long as only one kind of reality, the kind that can be (ill-)defined by mathematical means, and not reality as individuals see it, is permitted in mathematics education (Van der Blij, not dated). Very striking, in this regard, is de Lange's (1985) finding that it is often girls who point out the limitations of a mathematical model. Determining intervening factors in order to improve educational practice appears to be difficult.

Improving the System

In this approach, the content of mathematics education, or mathematics as a school subject in the context of society, is examined. Mathematics is not viewed as complete and immutable, but as a field which changes over time. Preferences in mathematics curricula relate to distinctions in the job market; distinctions in the job market relate to behavioural differences; behavioural differences relate to linguistic differences, etc. These relations are important for they tend to increase the stability of society.

In the remainder of this chapter I will focus on the third strategy outlined above. In the long-run, improvements at the level of the system, rather than at the level of the individual or the environment, may prove to be more effective in achieving a more permanent change in gender-related preferences for, and behaviour in, mathematics education.

The Language of Mathematics Education

In my view, mathematics education in the Netherlands, and in many other countries, is based on a certain dominant conception of mathematics: although the notion of mathematics as an open, increasing system seems to be accepted in theory, in practice mathematics itself is not a subject which is open to debate. In secondary school mathematics, the objectives are strictly prescribed, as if mathematics were just a set of facts that children can acquire or re-invent and use in different domains. 'Realistic contexts' are used as a means of rebuilding a pre-existing mathematical system. For example, 'the rules for differentiation of elementary functions are "discovered" or "re-invented" before the formal rules are finally given and proved . . . Finally, the students [embark] on the formal-analytical approach. The process of conceptual mathematization ends up with the differentiation rules, and their formal proofs' (de Lange, 1987, pp. 71–2).

I will distinguish between two basic conceptions of mathematics. The dominant conception focuses on closed forms of mathematics, the validity of which is based on assumptions or axioms that need not be justified by local contexts. For this reason I will call it 'expert-mathematics'. Inspired by the ideas of the Dutch mathematician, Brouwer (1907), the second conception of mathematics focuses on the design of various forms of mathematics which are justified by local needs and conditions. I call this second conception of mathematics 'user-mathematics'.

Expert-mathematics

Expert-mathematics is based on an absolutist view of knowledge which presupposes that we can distinguish objects in our world and represent them by abstractions. The relations between abstractions are described, isomorphic to the more 'concrete' relationships between the objects. Abstract relations are expressed as sentences in a formal language and constraints are put on which sentences, expressed as axioms, are relevant. Basing itself on abstraction, as it does, this language structures people's activities in this environment by focusing on measuring, computing and comparing.

Lakatos (1978) describes Euclidean geometry as the deduction of knowledge from trivial, true propositions (or axioms), on which the mathematical system is built. The deductive, or axiomatic, method is used to decide which sentences or propositions are accepted as belonging to the mathematical system. The legitimacy of the axiomatic method is beyond the realm of control of the user of mathematics, and lies instead in the presupposition of the universal knowledge of nature, seen as an external authority. It is assumed that it is possible to make statements about reality, by basing a mathematical system on axioms which stand to reason, and circumscribing the possible theorems by quality criteria such as coherence, consistency and completeness. As Lakatos (1977) says, 'in deductivist style, ... [m]athematics is presented as an ever-increasing set of eternal, immutable truths.'

The properties of 'expert-mathematics' guarantee 'a steadily accumulated body of knowledge, linear, hierarchical, dependable, reliable and value-free. Concepts do not develop, they are discovered' (Lerman, 1986, p. 71). 'Counter examples, refutations, criticism cannot possibly enter' (Lerman, 1986, p. 142).

The Restrictions on Pupils' Actions

Expert-mathematics implies a certain organization of the curriculum, which puts clear restrictions on pupils' actions, both in terms of what they are allowed to produce and what processes they are allowed to use to reach the desired results. Not all of these are necessary, and therefore, as Lakatos states, 'it has not yet been sufficiently realized that present mathematical and scientific education is a hot bed of authoritarianism and the worst enemy of independent and critical thought' (Lakatos, 1977, p. 142). As I stated above, the emphasis on knowledge-transmission creates a system of mathematics education in which students are taught to mimic experts. Children learn knowledge which is not necessarily useful to them in their (personal, or local) situations, or for solving their own problems. They are under the control of the teacher or the school textbook, outside authorities in matters which they feel are also their own, and this can lead to feelings of stress. Confrey (1985) argues that if rules, procedures and mathematics are seen as steady, certain and objective, children will perceive their own ideas, procedures, constructions and reflections as inferior and irrelevant. According to Confrey, this results in a negative attitude towards the subject, and generates apathy and aversion.

The dominance of expert-mathematics is also visible in Dutch realistic mathematics education. Although, in theory, children have freedom to explore structures in real-life situations, in practice they are allowed to find only prescribed structures in situations presented by their textbooks. Situations are chosen by educators because they provide students with suitable contexts for acquiring certain prescribed knowledge, rather than leaving children to construct or develop their own knowledge through exploration within the given context. The problems which are posed have implicit boundaries, which structure the possible solutions and the permitted activities. Because expert-mathematics is mainly used in science or technical domains, the contexts and problems which are chosen in education are mostly from those areas. Students whose 'reality' is not such a technical one and who respond in these contexts by avoiding expert-mathematics are viewed as mathematically untalented. The majority of these students are female. The school system, as it is presently organized, apparently induces avoidance of mathematics by women. To remedy this situation, we must abandon expert-mathematics and absolutist knowledge theory.

The Possibilities of User-Mathematics

User-mathematics, the second conception of mathematics mentioned above, provides an approach to mathematics education that holds much promise for increasing the participation of women in mathematics. It is related to a constructivist view of knowledge, the main proponent of which is Brouwer (1907). Mathematics, in Brouwer's view, does not deal with the representation of objects in reality. It must be seen as the product of human activity, as a mental construction, possibly shared by more than one person. According to Brouwer, the basic mathematical operation is the indication of a sequence of events in time. It is also the basis of action. 'These sequences were of a new kind: they were not given by a law (the automatic guarantee of infinite extension!) but by a choice process of the creating mathematician' (van Dalen, 1991, p. 9). Each person can make choices in a different way, and can order things in a way that is dependent of the needs and preferences of the individual. In this way, each person builds his or her own mathematical system. The main characteristic of user-mathematics is the legitimacy of the mathematics constructed. This is not necessarily bound to a certain 'objective reality', but may be subjected to a person's own needs and preferences in a particular situation. It is not necessary to stress external quality criteria, such as logical consistency or completeness. Every ordered series of events, produced by an individual, is seen as a certain mathematics, and the produced order is added to the series. In this way, external legitimacy is replaced by internal legitimacy. There is no need for any correspondence between relations and objects in the environment or in reality. 'I do not recognize as true, hence as mathematics, everything that can be written down in symbols according to certain rules, and conversely I can conceive mathematical truth which can never be fixed down in any system of formulas' (Brouwer *et al.*, 1937, p. 452). 'In this (Brouwerian) sense, mathematics is the overall systematic discipline dealing legitimately with *everything*' (van Dalen, 1991, p. 10). Mathematics gives expression to what people do, not to what they expect from their environment. Lakatos (1977) represents the same view when he states that mathematical activity produces mathematics. With a certain autonomy, mathematics develops as a living, growing organism from the initial activity. It emphasizes the way individuals deal with their environment, using its properties as constraints.

User-Mathematics in Education

User-mathematics construes a different type of pedagogy from the one required for expert-mathematics. In expert-mathematics, children are bound to see the regularities that teachers or curriculum designers have designated for them to see. In user-mathematics, by contrast, children have to find regularities in their own personal environment. They act as creative subjects; their activities are directed towards the preservation and extension of their own universes, rather than towards the design of an 'objective' universe. User-mathematics does not oblige students to use the same model, which is also not necessarily bound to a certain calculus. The users do not have to be identical, as expert-mathematics requires. User-mathematics does not select students on the basis of the possession of certain talents, or on the ability to solve calculable problems. Education of this kind is not restricted only to a select group of students, who recognize the prescribed world-view.

Several concerns about user-mathematics should be addressed. First, providing for more subjective constructions does not necessarily make pupils less effective in dealing with the world. Such constructions also take in what children want to do. Second, although children may impose linguistic constraints in their construction of mathematics, there is still the possibility of their communicating and sharing certain language elements. Children will be 'constructing their own knowledge, by comparing a new problem, idea, object, hypothesis against their existing experience, and conceptual system' (Lerman, 1986, p. 73). Shared constructs and shared knowledge are possible, but are not necessary. Both, however will fulfil the role of coordinating principles, rather than function as descriptions of what is.

Third, there is a serious danger of confusion in going only half-way. When teachers decide to adopt an approach which allows for the possibility of more personal and individual constructions, they often make problem-solving and problem-formulation their main forms of expression. The language of problem-solving then begins to dominate. All of mathematics is reformulated in its terms, but the underlying expert-mathematical nature of knowledge is maintained. The result is a confusion between what teachers take as a constructivist approach, and what they actually do.

User-Mathematics: A Promise for Girls?

In support of my claim, that user-mathematics will increase the participation of women in mathematics, I refer the reader to two situations in which students were able to choose their own methods of problem-solving using LOGO. In both cases, the authors, Sutherland and Hoyles (1988) and Turkle (1984), concluded that gender was no longer a reliable indicator of student motivation and performance. It is impossible for me to claim that every girl will have more affinity for mathematics after engaging in user-mathematics education. Neither can I claim that more equal numbers of boys and girls will choose to participate in more advanced mathematics, or in mathematically related domains, especially if these domains restrict themselves to current forms of mathematical modelling. But there is evidence that the choice and the exploratory mode that is characteristic of user-mathematics, and seldom present in expert-mathematics, is more attractive to girls. Apparently this is so, even in the area of assessment, for 'The creative constructive aspect of the take-home test, or the two-stage test (or essay), seems to especially appeal to girls. Most of the very best results came from girls' (de Lange, 1987, p. 261).

Examples of User-Mathematics in Education

There are few, if any, examples of the full implementation of user-mathematics in education. But there are examples which meet some of the criteria. These criteria are two-fold: first, the user is stimulated to a (personal) production of events; and second, users put those events in a certain order, including the order which has been produced, in order to deal with events and be more effective in the situation. To clarify this, I give two examples.

In a Dutch mathematics school textbook, a task is presented about holidays. In this (expert-mathematics) task, the students have to compare the costs of renting and driving two different cars over different distances. They are encouraged to be as clever as they can, but they are only allowed to operate within the very restricted area of tables and graphs. A user-mathematics approach, by contrast, would have students list all 'things', like goals, events, directions or ways of travelling that are essential in making their own holiday plans. Afterwards they are asked to order these 'things', to produce their own preferred way of travelling. This may lead to a comparison of the costs of renting and driving the two different cars, but it could also lead to the conclusion that some of the goals have to be changed in order to keep costs as low as possible, or that cost is of no importance at all. In user-mathematics, students are not judged on how they conform with standard rules of mathematical argument, but on the legitimacy of their own produced reasoning.

A recent project, aimed at high-school students in the Netherlands, has developed integrated mathematics activities which fit well with the ideas of user-mathematics. Students are allowed to choose between different kinds of tasks

(Obdeijn, Dalhoeven and Johannink, 1990), and the acceptance of work about cartoons, a light-game, and phases of the moon shows a break with current straight-jackets.

Conclusion

In this chapter I contend that gender differences in interest in mathematics may be difficult to change if we confine our strategies to changing the individuals or their environment. In many studies it has been demonstrated that approaches like these might change the landscape, but the differences remain. The same caution pertains to situations where one allows for some variability in the autonomy of those involved, but assumes that gender differences are dependent only on the way learning materials are delivered, for example, in a friendly or an authoritarian setting.

In contrast, I have proposed a third strategy based on the notion that gender differences in mathematics education are entirely system-induced. I have defined this strategy by describing one way of changing the 'language' of the educational system and hence the dominant conception of mathematics as currently presented in most schools. I have suggested that we should explore basing mathematics teaching more on a constructivist form of mathematics. I have argued that this will increase the variety of students involved in mathematics. I did not argue, however, that this strategy is sure to hit the nail on the head.

I want to conclude by mentioning some new opportunities. There is a widespread interest in constructivist approaches to education, and this language already has a fine and extensive implementation in constructivist mathematics education (see, for instance, Ernest, 1991; von Glasersfeld, 1991). There are many opportunities for applying the available instruments — but again, I have to stress, as I have done before, that it is easy to fail when one does not go the whole way, and also apply constructivist mathematics.

References

BLIJ, F. VAN DER (not dated) *owocABC*, Utrecht, Onderzoek Wiskunde-onderwijs en Onderwijs Computercentrum.

BROUWER, L.E.J. (1907) 'Over de grondslagen der Wiskunde', reprinted in DALEN, D. VAN. (1981) *L.E.J. Brouwer, Over de grondslagen der Wiskunde*, Amsterdam, Mathematisch Centrum, pp. 37–221.

BROUWER, L.E.J., EEDEN, F. VAN, GINNEKEN, J. VAN and MANNOURRY, G. (1937) 'Signific dialogues: Discussion on the formalistic method in significs', reprinted in HEYTING, A. (Ed) *L.E.J. Brouwer, Collected Works 1: Philosophy and Foundations of Mathematics*, Amsterdam, North Holland Publishing Company, pp. 447–52.

BURTON, L. (Ed) (1990) *Gender and Mathematics: An International Perspective*, London, Cassell.

CONFREY, J. (1985) 'Towards a framework of constructivist instruction', in STREEFLAND, L.

(Ed) *Proceedings of the Ninth International Conference for the Psychology of Mathematics Education*, Noordwijkerhout, pp. 477–83.

CONFREY, J. (1990) 'What constructivism implies for teaching', in DAVIS, R.B., MAHER, C.A. and NODDINGS, N. *Constructivist Views on the Teaching and Learning of Mathematics*, Journal for Research in Mathematics Education, Monograph no. 4, pp. 107–22.

DALEN, D. VAN (1991) 'Constructivism in mathematics', in DIJKUM, C. VAN and WALLNER, F. (Eds) *Constructive Realism in Discussion*, Amsterdam, Sokrates Science Publishers, pp. 6–17.

DEKKER, R., HERFST, P., TERWEL, J. and PLOEG, D. VAN DER (1985) *Interne differentiatie in heterogene brugklassen bij wiskunde: een empirisch-exploratief onderzoek naar de realisering en de resultaten van 'wiskunde voor iedereen' op een middenschool en een brede scholengemeenschap*, Den Haag, Instituut voor onderzoek van het onderwijs.

ERNEST, P. (1991) *The Philosophy of Mathematics Education*, London, Falmer Press.

FENNEMA, E. (Ed) (1985) 'Explaining sex-related differences in mathematics: Theoretical models', *Educational Studies in Mathematics*, 16, pp. 303–20.

GLASERSFELD, E. VON (Ed) (1991) *Radical Constructivism in Mathematics Education*, Dordrecht, Kluwer Academic Publishers.

HARDING, S. (1986) *The Science Question in Feminism*, Ithaca, Cornell University Press.

KELLER, E. FOX (1985) *Reflections on Gender and Science*, New Haven, CT, Yale University.

LAKATOS, I. (1977) *Proofs and Refutations: The Logic of Mathematical Discovery*, Cambridge, Cambridge University Press.

LAKATOS, I. (1978) *Mathematics, Science, and Epistemology, (vol. 2): Philosophical Papers*, London, Cambridge University Press.

LANGE, J. DE (1985) 'Nee', *De Nieuwe Wiskrant: Tijdschrift voor Nederlands Wiskundeonderwijs*, 2, 5, pp. 18–20.

LANGE, J. DE (1987) *Mathematics, Insight and Meaning: Teaching, Learning and Testing of Mathematics for the Life and Social Sciences*, Utrecht, Onderzoek Wiskunde-Onderwijs en Onderwijs Computercentrum.

LEDER, G. (1985) 'Sex-related differences in mathematics: An overview', in FENNEMA, E. (Ed) *Explaining Sex-Related Differences in Mathematics: Theoretical Models*, pp. 304–9.

LERMAN, S. (1986) *Alternative Views of the Nature of Mathematics and their Possible Influence on the Teaching of Mathematics*, London, King's College, University of London.

MEECE, J.L., PARSONS, J., KACZALA, C.M., GOFF, S.B. and FUTTERMAN, R. (1982) 'Sex differences in math achievement: Towards a model of academic choice', *Psychological Bulletin*, 91, 2, pp. 324–48.

MINISTERIE VAN ONDERWIJS EN WETENSCHAPPEN (1990) *Evaluatie voorlichtingscampagne Kies Exact*, Den Haag, Centrale Directie Voorlichting, Bibliotheek en Internationale Betrekkingen.

OBDEIJN, TH., DALHOEVEN, TH. and JOHANNINK, C. (1990) 'Schoolonderzoek in Oldenzaal', *Nieuwe Wiskrant: Tijdschrift voor Nederlands Wiskundeonderwijs, W 12-16 special 1*, 10, pp. 66–72.

PEDRO, J.D., WOLLEAT, P., FENNEMA, E. and BECKER, A. (1981) 'Election of high school mathematics by females and males: Attributions and attitudes', *American Educational Research Journal*, 18, 2, pp. 207–18.

SANGSTER, S. (1988) *Effect of Sex-Segregated Mathematics Classes on Students Attitudes, Achievement and Enrollment in Mathematics: A.Y. Jackson Secondary School, Year III*, Willowdale, Canada, North York Board of Education, ERIC, ED316420.

SHERMAN, J. (1981) 'Girls' and boys' enrollments in theoretical math courses: A longitudinal study', *Psychology of Woman Quarterly*, 5, 5, pp. 681–9.

SHERMAN, J. and FENNEMA, E. (1977) 'The study of mathematics among high school girls and boys: Related factors', *American Educational Research Journal*, 14, 2, pp. 159–68.

SUTHERLAND, R. and HOYLES, C. (1988) 'Gender perspectives on Logo programming in the mathematics curriculum', in HOYLES, C. (Ed) *Girls and Computers: General Issues and Case Studies of Logo in the Mathematics Classroom*, London, Bedford Way Papers, 34, pp. 40–63.

TURKLE, S. (1984) *The Second Self: Computers and the Human Spirit*, New York, Simon and Schuster.

VOS, E. (1988) 'Wiskunde leren in groepsverband', *Didaktief*, mei, pp. 10–12.

WALKERDINE, V. (1989) *Counting Girls Out*, London, Virago.

WITTE, M. (1992) 'Euclidean constraints in mathematics education', *Proceedings of the Sixteenth PME Conference*, Durham, NH, International Group for the Psychology of Mathematics Education, pp. 114–21.

WITTE, M. (1994) *Meisjes meegerekend. De constructie van wiskundige begaafdheid* (Girls count. The construction of mathematical competence), Amsterdam, Thesis Publishers.

Epilogue

We use the chapter of Nancy Shelley as our epilogue for it leads us to reflect how mathematics might look in the final phase of reform. Beginning from the assumption that we all carry with us a measure of reality which derives from the culture of western mathematics, Nancy Shelley stimulates us to imagine a mathematics which is not dominated by the experience of white men. She challenges the power of form and authority by adopting a non-traditional writing style. She calls to her aid T.S. Eliot's *Four Quartets* in seeking to summon those realities she wants to acknowledge. She asks fundamental questions about the status of knowledge and how it is culturally influenced, the epistemological status of issues in mathematics and mathematics teaching such as the role and function of absolutes, prediction, otherness, authority, objectivity. Describing her early experience of male domination in mathematics education, she makes a strong plea for challenging the relation of mathematics to war and calls for the necessity to fight for peace. She closes with a vision of a different mathematics, which grows out of a culture which has no pretension to dominance, which affirms humanness, and which relies on a culture of relationship, care and communion. According to Shelley the process by which it will grow is known, but the shape it will take remains to be determined.

Chapter 27

Mathematics: Beyond Good and Evil?

Nancy Shelley

Each of us carries with us, at any one time, a number of realities, and it is through the interlocking of those realities that we come to understand the world around us. Sometimes we consciously limit what we allow to impinge upon us, in order better to handle a particular reality pressing in its urgency. 'Comprehension', however, as Hannah Arendt[1] indicates, 'does not mean denying the outrageous, deducing the unprecedented from precedents, or explaining phenomena by such analogies and generalities that the impact of reality and the shock of experience are no longer felt'. It will come only through acknowledging other realities, both those of others and our own.

All of us carry with us a measure of reality which derives from the culture of western Mathematics. Since that culture permeates world views, values, ways of thinking, and its influence on attitudes is widespread, we need to face and understand realities which draw their nourishment from its bones.

I have used T.S. Eliot's *Four Quartets* to assist me in seeking to summon those realities we here would want to acknowledge. Just as he continually evokes images and thoughts which require us to ponder and think beyond the words used, I invite you to ponder similarly, allowing the ripples, which move out from the pebbles I drop into the pool of your listening, to stir the memory of your own experience.[2]

In this century, we have experienced the largest and most profound development in science known in recorded history, and the consequent technological explosion penetrates almost the entire globe. We are aware of an evolution in mathematical thinking of tremendous proportions, although less able either to detect the full extent of its present effect, or perceive its future ramifications. Changes wrought by moving from the absoluteness of space and the absoluteness of time to the concept of a space–time continuum have yet to impact upon our consciousness. We know of differences in mathematics education which affect more than the skills we employ. These dig deep into our individual philosophies and impinge upon our understanding of mathematics itself.

Time present and time past
Are both perhaps present in time future,
And time future contained in time past.[3]

In this century, too, the world is experiencing the largest displacement of people from their homes in recorded history; the highest number of people facing hunger, malnutrition and death from starvation; levels of violence within communities, between nations, within nations, in homes and globally which are unprecedented; rampageous destruction, degradation and pollution of the environment; species elimination on a scale not previously known. All of which co-exists with the greatest degree of universalist power ever exercised, yet which is accompanied by a wholesale disaffection with known political processes.

Time past and time future
Allow but a little consciousness.[4]

In the last decades of this century, it is possible to observe a burgeoning: in the re-awakening of cultural practice spawning new forms derived from old traditions; in the re-learning of human dependence upon the resources of the planet and interdependence with other life forms; in the re-appraisal of new shibboleths and old verities; in the re-discovery of the richness of understanding 'otherness' and a realization of the power of respectfulness; in a re-assurance that a polity will operate effectively only when it is centred round the people who form it.

At the still point of the turning world.
Neither flesh nor fleshless;
Neither from nor towards; at the still point,
there the dance is,
But neither arrest nor movement. . . .
Except for the point, the still point
There would be no dance, and there is
only the dance.[5]

In the last decades of this century we are witnessing the re-emergence of ethnic rivalries stemming from submerged identities: some being retrieved from centuries ago, some recasting a continual struggle, some being revived after half a century's incorporation. We can see the outbreaks of antagonism toward immigrant peoples reinvoking xenophobic fear, and we can see the regathering of opposition to repressive governments. Some of these manifestations exhibit new ideologies wrapped in old myths, some coalesce around an aggressive nationalism and many have become open markets for the arms trade. We can see the repercussions of economic ideologies bringing trauma through disintegration or from gaining ascendency. We can witness, too, the demands of indigenous people for recognition of their claims to land, their insistence that their dignity and their cultures be acknowledged, and, we can see some fragile attempts of the descendants of the invaders to appreciate their wisdom.

Words, after speech, reach
Into the silence. Only by the form, the pattern,
Can words or music reach
The stillness, as a Chinese jar still
Moves perpetually in its stillness.[6]

Developments of such magnitude, occurring concurrently, demand that any attempt at understanding, and locating our place in it all, must hold the reality of these developments in creative tension. Alongside ways of thinking lie ways of doing; alongside operating globally lies living locally; alongside dreams of domination lie nightmares of oppression; in, and through, and round about the exercise of power lies massive insecurity. And holding the reality of these rudiments in creative tension can be done only through telling a story.

I have been charged with this responsibility, and while any good story will have many stories within it, the prevailing one must needs be mine, carrying, as it does, the interlocking of my realities. I shall, too, etch in briefly some word pictures. The reality I derive from western Mathematics stems from teaching it for many years, being involved in teacher education and writing a thesis which was dealing with how different teaching methods affect the ways in which students come to acquire mathematical concepts. Other realities which I carry are grounded in the power of life-enhancing practices and adhere to my work, particularly over the past decade, in working, writing and taking action for peace. It will be in my story that I hope to encapsulate the things that need to be said under my title 'Mathematics: Beyond Good and Evil?'

> I asked the men, 'What are you carrying
> wrapped in that hammock, brothers?' And
> they answered, 'We carry a dead body, brother.'
> So I asked . . . 'Was he killed or did he die a
> natural death?' 'That is difficult to answer,
> brother. It seems more to have been
> a murder.' 'How was the man killed?
> With a knife or a bullet . . . ?' 'It was
> neither a knife nor a bullet; it was a
> much more perfect crime. One that leaves
> no sign.' 'Then how did they kill this man?'
> I asked and they calmly answered:
> 'This man was killed by hunger, brother'.[7]

Ability

Throughout the early 1970s, in a considerable number of schools I visited, I observed that not only large numbers of children were failing in Mathematics, but that they hated it. Teachers were dedicated, burdened and very puzzled. I saw, too, that many teachers relied upon ascertaining whether a child had ability, or not, to decide how that child would be taught.

I loved the subject and greatly enjoyed teaching it. I had always seen that it was my responsibility to find a way to engage a student in the thinking, and to do that I searched for the words, or the material, which brought forth his or her own experience. I could not believe it was possible to be definitive about whether a child was 'able' or 'not able'. And, for me, there is a qualitative difference between what we can know about objects and what we can know about another human being. German, for example, has the two verbs, *wissen* and *kennen*, which indicate this difference, whereas normal English usage has but one word, although the vestiges of distinction can be found in the language. I am indebted to Werner Pelz for the following:

> *Wissen* is both 'to know' and 'knowledge'. I know this, that and the other, know how to do this, manage that — *ich weiss*. . . . I may and must *wissen* how to run a certain machine. *Wissen* is certain or *gewiss*, yes, or no, binary. . . . *Wissen* gives manipulative power . . . It is a matter of skill, . . . precision.
>
> *Kennen* recognizes the other in his or her or its otherness, . . . [what] I can begin to *kennen*, has the character of . . . a confession of faith, of an *Anerkenntnis* or acknowledgment. . . . *Kennen* implies . . . sympathy, empathy, patience, openness, deference. It is close to love . . .[8] [Emphasis in original]

To 'know' a person is clearly distinguished from knowing an object. The merging of the two conceptual images permits the power and manipulation of humans, by other humans.

It is of interest here to consider the sensitivity with which Australian Aborigines approach coming to know about other human beings.

> Aboriginal social life is very public and personal privacy is ensured . . . through verbal privacy, particularly through an indirect style in much verbal interaction. . . . [I]n situations where Aboriginal people want to find out what they consider to be significant or certain information, they do not use direct questions. *Information is sought as part of a two-way exchange. Silence and waiting till people are ready, . . . are also central to Aboriginal ways of seeking any substantial information.*[9] [Emphasis in original]

> Modern . . . feeling[s] of isolation and powerlessness [are] increased . . . by the character which all . . . human relationships have assumed. The concrete relationship of one individual to another has lost its direct and human character and assumed a spirit of manipulation and instrumentality.[10]

Absolutes

When Newton formulated that,

> Absolute space, in its own nature, without regard to anything external, remains always similar and immovable,

and that

> Absolute true and mathematical time, of itself, and by its own nature, flows uniformly, without regard to anything external,[11]

he laid the foundation for the acceptance of the primacy and uniqueness of sequential time and for the replacement of the Aristotelian conception of space with one which henceforth became identified with the real space of the world. To acknowledge the absolute nature of space and time is to set them apart from humans — just as God had been set apart. To receive their absoluteness as true is to enable constructions built upon them to be separate from human experience.

As Polanyi put it:

> the mechanical properties of things alone were primary ... their other properties were derivative or secondary ... This too was a theoretical and objective view, in the sense of replacing the evidence of our senses by a formal space–time map that predicted the motions of material particles which were supposed to underlie all external experience.[12]

This 'formal space–time map' yielding its predictions of 'the motions of material particles' did not exempt humans in its majestic sweep 'to underlie all external experience'. As Popper noted:

> Most openminded men, and especially most scientists thought that in the end it would explain everything, including ... even living organisms.[13]

So it came about that humans, too, could be observed and measured; their actions predicted; they were likened to machines and the power of mechanism and physical determinism knew no bounds.

> *The knowledge imposes a pattern, and falsifies,*
> *For the pattern is new in every moment*
> *And every moment is a new and shocking*
> *Valuation of all we have been.*[14]

The teaching/learning process, particularly that relating to Mathematics, has also been affected by mechanism. When that process relies on seeing humans as being capable of being observed and measured by other humans through design of tests, through psychological practices, it is employing a mechanistic view of human beings, and then, more than the learning of Mathematics is affected. Questions of self-image are involved, and, of confidence which seeps through to other areas of students' lives.

> [If] I violate ... basic human dignity, I am being violent.[15]

From the noted Australian historian, Henry Reynolds, we learn that:

> Aboriginal self-confidence was not based solely on the mastery of particular skills but on the spiritual relationship with the land, the sense of belonging and responsibility . . .[16]

Prediction

Another element resides within our urge to 'know in advance'. The basis of prediction lies with our notion of causality and present ideas of cause and effect rely upon the acceptance of sequential time — given primacy by Newton. Aristotelian ideas of cause were teleological wherein 'the "nature" of anything is not its first but final condition' — toward which it progressed, toward perfection.[17] Instead of cause 'drawing you towards', it switched to 'driving you from behind' — the thing that drove then became more 'really' real than you.[18]

> *Here the impossible union*
> *Of spheres of existence is actual,*
> *Here the past and future*
> *Are conquered, and reconciled,*
> *Where action were otherwise movement*
> *Of that which is only moved*
> *And has in it no source of movement—*[19]

To come to know in ways that allow more and more to be foreseen, bestows a sense of power. It brings, as well, a sense of fear. As power from prediction increases, and when that power of intellect is transmuted through dominance to power political, 'tis then it seeks control. Only this year, in response to a law enacted in Washington State, the US Society of Psychologists found it necessary to make the public statement: 'You cannot predict future criminality.' Meanwhile there is little registering the accompanying insecurity which pervades the need to know events ahead of experiencing them.

Otherness

Throughout the period when I was writing the thesis, I was engaged in three particular programmes of mathematics teaching: one involved senior high-school students who had 'dropped out' of normal schooling, one involved women whose previous experience was that of having 'failed', and the third involved a pilot scheme with Aboriginal children who had been labelled 'unable'. Together these experiences increased my understanding of what was embedded in the subject called Mathematics and what was built into the concept of failure in the subject.

In each of these three situations, I placed a priority upon establishing an atmosphere of trust, recognizing the other in his or her otherness. By an 'opening up' and a 'drawing in'[20] to an active experience of relation one to another, and of relation of each to the 'matter' in mathematics, it was possible to develop a new understanding of it. As the criteria for 'correctness' were re-established within each of the participants, that is, as they learned to take responsibility for decision-making, a process of learning took place, and mathematics became a meaningful exercise. The association with which they came to know affected what it was they learnt. Mathematics began to reveal some of its richness and became enjoyable as it offered new ways of thinking.

Authority

In my work in teacher education, I had noted that the diversity which the students displayed presented a problem for the school and for the teachers, and indeed, it had become a pedagogical nuisance. At the same time, I came to realize that a strong factor which affected teachers was the authority the subject held for them, while a common perception of Mathematics in society was its rock-like irrefutability.

For me, diversity is something to celebrate as it brings into any experience the unexpected, which constantly reminds me of the ever-renewing capacity of living beings. Besides, when diversity appears a problem, uniformity is lurking in the wings!

Desiccation of the world of sense,
Evacuation of the world of fantasy,
Inoperancy of the world of spirit;
. . . while the world moves
In appetency, on its metalled ways
Of time past and time future.[21]

The world of spirit was far from inoperative throughout the eighteenth and most of the nineteenth centuries, nor had the world of fantasy been evacuated, as mathematical insights came and ideas were born, as one writer described it, 'in a veritable orgy of intuitive guesswork'. The 'ecstasy of progress' opened up 'a mathematical world of immense riches'.[22]

It was not until the end of the nineteenth century that there was a return to the classical Greek ideal of precision and rigorous proof: to the employment of 'logically precise reasoning starting from clear definitions and non-contradictory "evident" axioms'.[23] With the arduous work of the Logicists, working on the several 'pieces', we had a shaping of those 'immense riches', and all others that followed them, in an attempt to fashion a whole — a hierarchical whole which, it was hoped, would exhibit the same characteristics: consistency throughout, using argument based on logic, and, upon completion, the acknowledgment of validity. The result has been a very substantial, indeed formidable, edifice. Yet, the sum of innumerable distinct fragments of validity does not constitute a verdict of truth.

If the logic employed has, concurrently, become the basis for public argument and accepted rationality — as it has — it also allows the whole to assume an authority which is daunting and a visage of irrefutability which exudes great power. It was the *authority* of Aristotle which played a large part in the restricting of human thought for centuries. 'Tis this that most people encounter when introduced to western Mathematics. 'Tis this which has earned its description as an objective body of knowledge. 'Tis this which further disconnects it from human experience.

Objectivity

> People now . . . tend to trust their own experience, their eyes and ears, less than a system which by virtue of its consistency is more rational than reality.[24]

In the opinion of many mathematicians this objectivity constitutes a strength; for most learners it is an impediment; for many teachers it is inhibiting and for much teaching/learning it becomes an imposition. For the subject itself, it is damaging — it separates it from its roots, falsifies its history, and has dramatically influenced the direction its development has taken. Meanwhile, it ignores how mathematical ideas originate and limits mathematical thinking by channelling it in selected paths.

Also, in its wake it brings a number of characteristics. It puts a high value on consistency and logic; validity is prized. It is hierarchical. Through systematizing how Mathematics is to be viewed, the basis for decision-making has been taken out of the hands of teachers, of students, and, indeed, of all humans who come in contact with it; 'correctness' hereafter resides within the system — it becomes the authority by which everything is decided.

Any statement made out of objective reality comes across as a command; it dictates; certain conclusions are 'necessary'. The system itself is now one of static permanence; its validity has become truth, and, as Hannah Arendt observed:

> the kind of truth derived from the paradigm of truth dependent upon the logical argument which informs western thought, is essentially coercive and, of its nature silences opinion and tries to impose uniformity.[25]

Through its logic this Mathematics has become coercive and imposes uniformity, for

> . . . the truth of a system . . . [comes] to be a threat to freedom . . . because it allows no room for human diversity.[26]

Acceptance of this visage requires an imposition also, for it prescribes what shall be learnt, and, that does not derive in any way from one's experience; it requires compliance.

> The more one conforms to anonymous authorities,
> the more one adopts a self which is not one's own,
> then, the more powerless one feels and
> the more one is forced to conform.[27]

If we remove humans and the earth from relationship, we turn both into objects to be acted upon; this is violence. This Mathematics holds over the heads of prospective learners impending decisions concerning their ability, particularly relating to how logical they are. It dominates. It seeks obedience. It seeks control. The subject is then seen to be more important than a student's engagement in it.

This story of a Korean woman expresses it clearly:

> The major focus in the curriculum was learning mathematics and English and other foreign languages . . . Good grades automatically led us to good colleges. We frantically memorized the foreign vocabulary and mathematical formulas like robots so we could get good grades. It was a long and lonely adolescence.[28]

Male Domain

I, too, was beginning to feel lonely as I endeavoured to tease out and put flesh on the bones of my thinking, for I could not find mathematics people who knew what I was saying. Then, in 1976, I came to my first ICME, in Karlsruhe, and, thinking to test the waters, I offered a 'Short Communication'. Two things stand out for me from that Congress.

At a special session which was hastily arranged, along with others, I was invited to speak. There ensued a dialogue with some of the participants which was both lively and affirming. That dialogue was most important for me because, instead of feeling isolated, I learned there were other people across the globe who were exploring similar paths.

The second thing that stands out for me from that Congress was the extraordinary feeling I had that it was totally male-dominated, despite the presence of a large number of women participants. I experienced very strong feelings of being pushed down and being held there, without any opportunity to comment on what was happening. I checked out my impressions with some Australian colleagues — both female and male — and found that they, too, felt it. So we decided to do something about it and called a meeting of women 'to talk about ICME'.

About seventy-five people came, both women and men, and out of that meeting IOWME, the International Organization of Women and Mathematics Education, was born. IOWME has affected the format of each ICME since, helped to bring the question of women and Mathematics into the arena, and now has branches in more than forty countries.

However, the path by which we moved from first calling that meeting to the establishment of IOWME and the acceptance of its first resolution, which we sought to present to the final plenary, was not a smooth one. Some women were actively discouraged from attending the meeting by male colleagues from their own country. Attempts were made to break up our meeting. We had to deal — and I choose my words carefully — with anger, alarm, obstruction and public derision.

Of course, I was aware that Mathematics was considered a male domain, I had experienced non-acceptance at times, and knew that most girls did not do Mathematics. However, here was a direct experience of being oppressed by the authority of a large number of men who, very visibly, exercised control over proceedings, and apparently over what was acceptable as to how mathematics education should be developed. I had not before experienced the power of ownership which men, eminent in their field, sought to exercise.

From time to time, particular groups of people with common interests come together with the specific purpose of furthering the development of an understanding of the world along certain particular paths. Western Mathematics is the product of the work of such a group, and it constitutes a very elaborate model of reality built up over several centuries.

As Ursula Franklin noted:

> Models and analogies are always needed for communication, and in order to be useful tools for discussion, models and metaphors need to be based on shared and commonly understood experiences.[29]

Western Mathematics has moved away from 'commonly understood experiences'. Its beginnings lie in the common experience and culture of European men, and constitute a model which they found enabled them to interpret the world. Much has been built upon those beginnings; it has, indeed, become an extremely elaborate construction. It remains, however, a model, albeit a very powerful one.

If a map is mistaken for the territory itself, unreality becomes its monarch. If this model — from western Mathematics — is confused with reality itself, validity is translated into truth, what is one model only becomes the only reality — and a static one — and the intellectual power which the model bestows gives way to rigidity requiring defence: to **convince** assumes importance and **justification**, compelling.[30]

As Barbara McClintock recalled:

> I couldn't tell other people at the time because it was against the 'scientific method'. . . . we just hadn't touched on this kind of knowledge . . . [and it is] very different from the knowledge we call the only way.[31]

Go, go, go, said the bird: human kind
Cannot bear very much reality.[32]

Now any model of reality has power for those who construct it, and a model shared becomes the basis of a culture, centred round the reality which the model affords, and bestowing its power. The construction of western Mathematics arising out of the culture of European men, and, having, as it does, their 'historically determined values built into it', as Mike Cooley expressed it, it must needs be a male domain.[33] It is not surprising, therefore, in the presence of the heirs and architects of that model, to experience a sense of ownership.

I said to my soul, be still, and wait without hope
For hope would be hope for the wrong thing; wait without love
For love would be love of the wrong thing; there is yet faith
But the faith and the love and the hope are all in the waiting.
Wait without thought, for you are not ready for thought:
So the darkness shall be the light, and the stillness the dancing.[34]

Peace and War

What a bewildering world . . .
They feed people with words,
the pigs with choice potatoes.[35]

Each day, about 300 km^2 of tropical forests are cut down,
and, each day, around 160 km^2 of land turns to desert.

We must be still and still moving
Into another intensity
For a further union, a deeper communion
Through the dark cold and the empty desolation,
. . . *In my end is my beginning.*[36]

The thesis was duly completed and it led me on to articulate more clearly what I see as central to the teaching/learning process, to develop new methods of teacher education and to run workshops for practising teachers.

Then, in the early 80s, I made the decision to give up paid employment in order to work full-time for peace. In doing this I was giving precedence to a reality which had become more and more important to me. My commitment to peace did not arise out of fear of the possibility of war, but is grounded in my convincement of the power of life-enhancing practices.

In most western minds throughout the 80s there was a preoccupation with the possibility of annihilation which came out of the heightened confrontation between the western powers and the Soviet bloc, made feasible by the enormity of opposing military juggernauts.

The number of people dying [in the world]
as a result of malnutrition is equivalent to
dropping a Hiroshima bomb *every three days*.
[Emphasis in original][37]

The engineering of the world's annihilation goes hand in hand
with the engineering of the world's consent.[38]

When massive numbers of people withdrew their consent, the superpowers backed away from the brink.

Towards the end of the 80s came a *rapprochement* between the superpowers, and as *glasnost* affected the thinking in both east and west, the visage of 'the enemy' dissolved. This brought much rejoicing which swelled in crescendo with the crumbling of the Berlin Wall. There was, however, a murmur of insecurity.

People change, and smile: but the agony abides.[39]

When the crumbling became the symbol of a vaster internal fragmentation that sought its *perestroika*, the murmur of unease became persistent. There was little understanding that strategists, the military, and the industrial complexes, which depend upon governments' subsidization, were working from different premises.

Militarism has proven to have a particularly powerful
ideological pull.[40]

Where is there an end of it, the soundless wailing,
The silent withering of autumn flowers
Dropping their petals and remaining motionless;
Where is there an end to the drifting wreckage—[41]

During the first half of 1990, the Pentagon devised its new military strategy of mid-intensity conflict to replace that of high-intensity previously directed toward the Soviet Union.

There is no end, but addition—[42]

In the first half of 1991 we witnessed the Gulf War, which marshalled the combined forces of many countries. While the Iraqi people faced the reality of being bombarded, on TV screens throughout the world, viewers were bombarded with a virtual-reality display of the 'theatre' of war. Weapons took on human attributes: 'smart' weapons had 'eyes', computers 'brains' which 'made decisions', and it was military targets and missiles which were 'killed'. Human beings acquired the characteristics of objects: if they were military personnel they 'absorbed munitions'; if civilian, they were not hurt, maimed or killed but recorded as part of 'collateral damage'.[43]

There is no end of it, the voiceless wailing,
No end to the withering of withered flowers,
To the movement of pain that is painless and
motionless,
To the drift of the sea and the drifting wreckage . . .[44]

The repercussions of that war will continue to reverberate for many years and in many ways. Few reactions will surpass the effect of the dehumanizing that took place: through people's senses, through being an audience — assisted by that extraordinary switch of language — through the total exclusion of reference to maiming or killing of human beings and through the absence of outrage. The desensitization will have found its way into the crevices of people's minds.

the excessive power of a few leads to the dehumanization of all.[45]

. . . we come to discover that the moments of agony
. . . are . . . permanent
With such permanence as time has.[46]

At a NATO meeting in November 1991, the murmur was openly articulated when George Bush announced that since the loss of 'an enemy' we now have 'insecurity' and 'instability'. From the perspective of power, it appears that security and stability derive from having 'an enemy'.

Do not let me hear
Of the wisdom of old men, but rather of their folly,
Their fear of fear—[47]

The values embraced in militarist ideology have taken
on their own life and their own laws.[48]

It seems, as one becomes older,
That the past has another pattern, and ceases to
be a mere sequence —
Or even development:[49]

From my past I began to see another pattern. Through my research into matters relating to war and peace, I learned that by far the greatest number of mathematicians are engaged in military research and development. In consequence, the single greatest area of mathematics work, today, is problem-solving for the military; the foundations of military science are underpinned by western Mathematics.

Mathematics, of course, can be put to innumerable uses; some of those uses are disowned by mathematicians, some are manifestly invalid. Here, however, there are different matters to consider. First, since the military attracts, and funds, the majority, and those designated the brightest, of today's mathematicians, we need to realize that the major development of mathematics is now channelled in very selected ways.

Another matter germane to us here is why so many mathematicians fit readily into a particular type of research? What makes it conducive for one trained in mathematical thinking to be at ease within that cast found in a military culture?

We live in a world where western culture is dominant; in a world where decisions made are described as logical and rational; where power is exercised through hierarchies; where authority has clout; where dominance is the norm; where power is exercised by the few over the many; where order is imposed by 'knowledgeable' people; where power is enforced by military might.

Western Mathematics plays a dominant role in western culture. People with mathematics qualifications are highly respected; they enjoy status; their work is valued as a necessary and integral part of society. The logic which permeates western Mathematics, having been the basis of accepted rationality in western thinking for centuries, is now pervading the globe. Dealing with problems, it is common practice to redefine them in such a way as to facilitate their mathematization.[50] The culture of western Mathematics derives its strength from the dominance of its model of reality. This culture is now a dominant culture in the world, and, as with any dominant culture, it lays claim to universality, to actuality and to uniqueness. Western Mathematics is value-laden.

> Every relationship of domination . . . is by definition violent, whether or not the violence is expressed by drastic means. In such a relationship, dominator and dominated alike are reduced to things — the former dehumanized by an excess of power, the latter by lack of it.[51]

We live, too, in a world where the power of the military is great: where military 'solutions' are seen to be most effective, and the carrying out of orders most efficient; where training develops a particular reality which becomes mandatory; where problems are 'refined' to eliminate ambiguity; where the precision of military decision-making is admired amid the complexities of human issues; where training of personnel involves 'killing the other' in them; where the military is a major contributor to land degradation; where the military is a primary source of environmental poisoning; where it is recognized that an efficient killer is one who no longer contemplates the humanness of the enemy.

> *Here between the hither and the farther shore*
> *While time is withdrawn, consider the future*
> *And the past with an equal mind.*[52]

Concurrently, in our world, increasingly, we can find more and more people: living frugally in order to be more responsible; regaining dignity through creative art; designing and constructing energy-efficient buildings; establishing community; developing relationship with the land; devising practical schemes of assistance; restoring ecosystems; renewing relationships within small groups; repairing environmental damage; networking with others locally, nationally and internationally.

Time past and time future
What might have been and what has been
Point to one end, which is always present.[53]

There is, too, the possibility of a different mathematics which grows out of a culture which has no pretensions to dominance; which can yet engage a learner creatively; which affirms humanness; and which attends human needs and will embrace earthy things. A culture of relationship, of care, of communion. The process by which it will grow is known, the shape it will take remains to be made.

To become renewed, transfigured, in another pattern.[54]

Today, too, we live in a culture of violence. A culture: where to dominate is to be strong; where power is maintained by violence; where to conquer is to stand tall; where some lives are counted more valuable than others; where decisions made to expend resources upon weapons are called rational while millions starve; where torture is frequently applied; where there is violence against the 'other', against the poor, the hungry, women, the young, racial violence; where human dignity is constantly violated; where competition is promoted and to win is essential, the loser despised; where conditions are not negotiated, but imposed; and, where the violence of repression, of exploitation, of unemployment, of oppression, and destruction of the earth are widespread.

What we call the beginning is often the end
And to make an end is to make a beginning.
The end is where we start from.[55]

Competing realities of such magnitude, occurring concurrently, require that we constantly hold them in creative tension, if we want to attempt understanding. Each of us will take our place as we see fit; different realities will have different pulls upon us.

To postpone or evade the decision is still to decide.
To hide the matter is to decide. To compromise is to
decide. There is no escape from this decision . . .[56]

Is Mathematics beyond good and evil?

In order to arrive at what you do not know
You must go by a way which is the way of ignorance.
In order to possess what you do not possess
You must go by the way of dispossession.
In order to arrive at what you are not
You must go through the way in which you are not.
And what you do not know is the only thing you know
And what you own is what you do not own
And where you are is where you are not.[57]

Notes

1 Arendt, H. (1973) *The Origins of Totalitarianism*, New York, Harcourt Brace Jovanovich, page viii.
2 'Mathematics: Beyond Good and Evil?' is written to appeal to more human characteristics than logic. One of those is hearing. In order, therefore, to hear its rhythm, the author recommends that the chapter be read aloud. The author was invited by the organizers of ICME-7 to present this paper as one of the Lectures at the ICME-7 Congress, in Québec City, in August 1992. It was well received. The Committee responsible for publication of those lectures declined to publish it. So the editors of this book, which is based on papers presented in the IOWME sessions at ICME-7, decided to include it.
3 Eliot, T.S. (Mcmlviii) *Four Quartets*, London, Faber and Faber, page 7.
4 *Ibid*, page 10.
5 *Ibid*, page 9.
6 *Ibid*, page 12.
7 Bennett, J. with George, S. (1987) *The Hunger Machine: The Politics of Food*, New York, Polity Press, page 21.
8 Pelz, W. (1974) *The Scope of Understanding in Sociology: Towards a More Radical Orientation in the Social and Humanistic Sciences*, London, Routledge and Kegan Paul, pages 80–81.
9 Eades, D. (1992) *Aboriginal English and the Law*, Continuing Legal Education Department of the Queensland Law Society Incorporated, page 28.
10 Fromm, E. (1960) *Fear of Freedom*, London, Routledge and Kegan Paul, page 102.
11 Newton, I. (1970) 'Philosophiae Naturalis Principia Mathematica', in Bas, C. van Fraasen, *An Introduction to the Philosophy of Time and Space*, New York, Random House, page 23.
12 Polanyi, M. (1973) *Personal Knowledge: Towards a Post-Critical Philosophy*, London, Routledge and Kegan Paul, page 8.
13 Popper, K. (1972) *Objective Knowledge: An Evolutionary Approach*, Oxford, Clarendon Press, page 211.
14 Eliot, *op cit*, page 18.
15 Zaru, J. (1986) 'May God's peace, mercy, and blessings be with you', in Eck, D.L. and Devaki, J. (Eds) *Speaking of Faith: Cross-Cultural Perspectives on Women, Religion and Change*, London, The Women's Press, page 54.
16 Reynolds, H. (1982) *The Other Side of the Frontier: Aboriginal Resistance to the European Invasion of Australia*, Ringwood, Penguin, pages 58–59.
17 Aristotle (1964) *The Politics*, translated by T.A. Sinclair, Harmondsworth, Penguin, in the Introduction, page 28.
18 Heller, E. (1961) *The Disinherited Mind*, Harmondsworth, Penguin, page 186.
19 Eliot, *op cit*, page 33.
20 Buber, M. (1971) *Between Man and Man*, London, Collins, page 118.
21 Eliot, *op cit*, page 11.
22 Courant, R. and Robbins, H. (1948) *What is Mathematics?: An Elementary Approach to Ideas and Methods*, London, Oxford University Press, page xvi.
23 *Ibid*, page xvi.
24 Canovan, M. (1974) *The Political Thought of Hannah Arendt*, London, Methuen, page 21.

25 *Ibid*, page 114.
26 *Ibid*, page 114.
27 FROMM, E. quoted in FREIRE, P. (1973) *Education For Critical Consciousness*, New York, Seabury Press, pages 6–7.
28 CHUNG, H.K. (1990) *Struggle to be the Sun Again*, Orbis Books, page 2.
29 FRANKLIN, U. (1990) *The Real World of Technology*, Massey Lecture Series, Montréal, CBC Enterprises, page 26.
30 Adapted from MURA, R. (1991) *Searching For Subjectivity in the World of the Sciences: Feminist Viewpoints*, CRIAW/ICREF, page 52.
31 MCCLINTOCK, B. quoted in KELLER, E.F. (1983) *A Feeling for the Organism: The Life and Work of Barbara McClintock*, New York, W.H. Freeman and Company, page 203.
32 ELIOT, *op cit*, page 8.
33 COOLEY, M. (1980) *Architect or Bee? The Human/Technology Relationship*, Sydney, TransNational Co-operative Limited, page 43.
34 ELIOT, *op cit*, page 19.
35 BENNET, *op cit*, page 37.
36 ELIOT, *op cit*, pages 22–23.
37 BENNET, *op cit*, page 12.
38 OTERO, C.P. in the Introduction to Chomsky, Noam. (1981) *Radical Priorities*, Montréal, Black Rose Books, page 13.
39 ELIOT, *op cit*, page 29.
40 SMITH, D. and SMITH, R. (1983) *The Economics of Militarism*, London, Pluto Press, page 50.
41 ELIOT, *op cit*, page 27.
42 *Ibid*, page 27.
43 Adapted from COHN, C.E. (1991) 'Decoding Military Newspeak', in *Ms.: The World of Women*, 1, 5, page 88.
44 ELIOT, *op cit*, page 28.
45 FREIRE, *op cit*, page 10.
46 ELIOT, *op cit*, page 29.
47 *Ibid*, page 18.
48 SMITH and SMITH, *op cit*, page 52.
49 ELIOT, *op cit*, page 28.
50 MURA, *op cit*, page 54.
51 FREIRE, *op cit*, footnote on page 10.
52 ELIOT, *op cit*, pages 30–31.
53 *Ibid*, page 8.
54 *Ibid*, page 41.
55 *Ibid*, page 42.
56 ZARU, op cit, page 55.
57 ELIOT, *op cit*, page 20.

References

ARENDT, H. (1973) *The Origins of Totalitarianism*, New York, Harcourt Brace Jovanovich.

ARISTOTLE (1964) *The Politics*, translated by Sinclair, T.A., Harmondsworth, Penguin, in the Introduction.

BENNET, J. with GEORGE, S. (1987) *The Hunger Machine: The Politics of Food*, New York, Polity Press.

BUBER, M. (1971) *Between Man and Man*, London, Collins.

CANOVAN, M. (1974) *The Political Thought of Hannah Arendt*, London, Methuen.

CHUNG, H.K. (1990) *Struggle to be the Sun Again*, Orbis Books.

COHN, C.E. (1991) 'Decoding Military Newspeak', in *Ms.: The World Of Women*, 1, 5.

COOLEY, M. (1980) *Architect or Bee? The Human/Technology Relationship*, Sydney, Trans-National Co-operative Limited.

COURANT, R. and ROBBINS, H. (1948) *What is Mathematics?: An Elementary Approach to Ideas and Methods*, London, Oxford University Press.

EADES, D. (1992) *Aboriginal English and the Law*, Continuing Legal Education Department of the Queensland Law Society Incorporated.

ELIOT, T.S. (Mcmlviii) *Four Quartets*, London, Faber and Faber.

FRANKLIN, U. (1990) *The Real World of Technology*, Massey Lecture Series, Montréal, CBC Enterprises.

FROMM, E. (1960) *Fear of Freedom*, London, Routledge and Kegan Paul.

FROMM, E. quoted in FREIRE, P. (1973) *Education For Critical Consciousness*, New York, Seabury Press.

HELLER, E. (1961) *The Disinherited Mind*, Harmondsworth, Penguin.

MCCLINTOCK, B. quoted in KELLER, E.F. (1983) *A Feeling for the Organism: The Life and Work of Barbara McClintock*, New York, W.H. Freeman and Company.

MURA, R. (1991) *Searching For Subjectivity in the World of the Sciences: Feminist Viewpoints*, CRIAW/ICREF.

NEWTON, I. (1970) 'Philosophiae Naturalis Principia Mathematica', in VAN FRAASEN, BAS C., *An Introduction to the Philosophy of Time and Space*, New York, Random House.

OTERO, C.P. in the Introduction to CHOMSKY, N. (1981) *Radical Priorities*, Montréal, Black Rose Books.

PELZ, W. (1974) *The Scope of Understanding in Sociology: Towards a More Radical Orientation in the Social and Humanistic Sciences*, London, Routledge and Kegan Paul.

POLANYI, M. (1973) *Personal Knowledge: Towards a Post-Critical Philosophy*, London, Routledge and Kegan Paul.

POPPER, K. (1972) *Objective Knowledge: An Evolutionary Approach*, Oxford, Clarendon Press.

REYNOLDS, H. (1982) *The Other Side of the Frontier: Aboriginal Resistance to the European Invasion of Australia*, Ringwood, Penguin.

SMITH, D. and SMITH, R. (1983) *The Economics of Militarism*, London, Pluto Press.

ZARU, J. (1986) 'May God's peace, mercy, and blessings be with you', in ECK, D.L. and DEVAKI, J. (Eds) *Speaking of Faith: Cross-Cultural Perspectives on Women, Religion and Change*, London, The Women's Press.

Notes on Contributors

Joanne Rossi Becker has a doctorate in mathematics education from the University of Maryland and is a professor of mathematics at San José State University. Much of her research has been in the area of gender and mathematics, including classroom-interaction studies and interviews with graduate students in the mathematical sciences. She has been active in professional groups, serving as chair of Women and Mathematics Education, and of a special interest group of the American Educational Research Association on Research on Women and Education, and coordinating the IOWME sessions at ICME-5 in Australia.

Paul Brandon has a doctorate in educational psychology and is an assistant professor of educational evaluation at the University of Hawai'i at Mänoa. He has studied educational achievement among Asian and Pacific Americans, including gender differences in mathematics achievement among Hawai'i's four major ethnic groups and in the educational attainment of Asian Americans nationwide.

Leone Burton is a professor of education (mathematics and science) at the University of Birmingham. She has been closely involved with mathematics teacher education in England and has experience teaching students of all ages from the very young to adults. Her research interests began with problem-solving but developed into a parallel concern with social-justice issues and the nature of mathematics. Like many others with a professional interest in mathematics, music is also a passion. She plays (more or less badly) and enjoys listening particularly to baroque chamber music and to modern jazz.

Françoise Delon works in the field of mathematical logic. She is 'directrice de Récherche' at the 'Centre National de la Récherche Scientifique', an affiliate of the University of Paris. She has been active in the association 'Femmes et Mathématiques' since its inception in 1987, and was its president from 1989 to 1990.

Sharleen D. Forbes, in addition to being the mother of four children, has had over twelve years of secondary-school and university experience in mathematics education, with a focus on the teaching of statistics. She has been involved in mathematics research for the last decade, with particular areas of interest being 'second-chance' education in mathematics, the evaluation of gender and ethnic differences in mathematics performance, and the assessment of mathematics learning.

Lynn Friedman has masters degrees in mathematics and in statistics and obtained her doctorate in mathematics education from the University of Chicago. Currently, she is an assistant professor in the Department of Educational Psychology at the University of Minnesota where she specializes in meta-analysis and cognitive gender differences.

Olive Fullerton has a doctorate in education from the University of Toronto and is a professor in the Faculty of Education at York University where she teaches both undergraduate and graduate students. Her research interests include dysfunctional attitudes around mathematics, especially in practising teachers and how those attitudes impact on young children, how young children learn mathematics, and especially the role of oral language in learning mathematics.

Barbro Grevholm has a masters degree in mathematics, theoretical physics, physics and mathematical statistics. She has many years of experience teaching in the upper-secondary school system as well as lecturing at the Mathematics Institution in Lund. Since 1985 she has been teaching part-time at the School of Education in Malmo.

Saleha Naghmi Habibullah was born in Pakistan and has lived for most of her life in Lahore. She has masters degrees in statistics from both the University of the Punjab and the University of Toronto. She has taught at Kinnaird College for Women in Lahore since 1979 where she is currently head of the statistics department. Her primary area of interest is in the development of statistical education at the undergraduate level in Pakistan.

Mary Harris became interested in problems associated with learning mathematics after a number of part-time jobs while bringing up her mentally disabled daughter. After obtaining a degree in education, she worked in mathematics education with 'low attainers'. Currently she is a visiting fellow in the Department of Mathematics, Statistics and Computing at the University of London Institute of Education.

Pat Hiddleston is an associate professor of mathematics education at the University of Malawi where she is also head of the Department of Curriculum and Teaching Studies. Her research interests are in gender issues in mathematics education and in the comparison of attitudes to mathematics between developing and developed countries.

Terry Ann Higa has a professional diploma in elementary education, and masters degrees in secondary-education curriculum and instruction, and in educational psychology. She is a graduate assistant for the Curriculum Research and Development Group at the College of Education in the University of Hawai'i, Mänoa. She recently completed a study of the validity of an instrument on the cognitive approaches of medical-school students.

Joanna Higgins is a doctoral candidate at Victoria University of Wellington and teaches mathematics education at Wellington College of Education/Te Whanau o Ako Pai ki Te Upoko o Te Ika, New Zealand/Aotearoa. She has taught in elementary schools and is currently working with teachers on developing their classroom mathematics programmes.

Betty Johnston is a lecturer in numeracy in the School of Adult and Language Education at the University of Technology in Sydney, Australia. She has taught mathematics to students of all ages, from 5 to 85 years of age, in a variety of countries — New Zealand, Canada, England, Ghana, Papua New Guinea and Australia. She has also spent many years teaching teachers of those students. Her particular interest is in what might be called critical numeracy.

Cathie Jordan holds a doctorate in anthropology and is a researcher and programme developer with the Early Education Division of Kamehameha Schools/Bishop Estate in Honolulu. Her current research is on factors affecting family utilization of preschool educational programmes and on the incorporation of Hawaiian values into educational programmes for native Hawaiian children.

Gurcharn Singh Kaeley was born, raised, and educated in India. His teaching career began in Kenya in 1963 where he served as mathematics instructor at the University of Nairobi for several years. Since then, he worked as a mathematics coordinator in the Department of Extension Studies, University of Papua New Guinea where he did research on the teaching of mathematics to mature students with a focus on gender differences. Currently he is working with the University of Chicago School Mathematics Project (secondary component) as director of evaluations.

Gabriele Kaiser holds a masters degree as a teacher for mathematics and humanities and has taught in schools for a few years. She has completed her doctorate in mathematics education and is currently working at the University of Kassel doing empirical research comparing mathematics education in England and Germany. Co-founder, with Christine Keitel and Cornelia Niederdrenk-Felgner, of the German IOWME section, she now serves as its national coordinator. Jointly with Pat Rogers she organized the IOWME sessions at ICME-7. She has a little daughter and in her spare time she loves to play tennis and do aerobic exercises.

Berinderjeet Kaur is a mathematics-education lecturer in the School of Science in the National Institute of Education, Nanyang Technological University, Singapore. She has a masters degree from the University of Nottingham and is currently working on her doctorate at Monash University, Australia. She is the national coordinator of IOWME for Singapore and her research interests are in the areas of problems in mathematics and in gender issues.

Hélène Kayler is a professor of mathematics education in the Département de Mathématiques of the University of Québec in Montréal. She has a masters degree in mathematics, specializing in mathematics education. Currently, she is involved in elementary-teacher education and doing research on the learning of mathematics with children and adults. She was president of MOIFEM, the Québec section of IOWME, from 1990 to 1992.

Louise Lafortune is a professor and researcher at the Collège André Laurendeau and a researcher at the Centre Interdisciplinaire de Récherche sur l'Apprentissage et le Dévèloppement en Education (CIRADE) at the University of Québec in Montréal. For her masters degree, she focused on the history of women mathematicians and for her doctorate, on the affective domain in mathematics education. Her fields of interest are the role of affectivity and metacognition in the learning process, equity in pedagogy, and qualitative research methods. She was founding president of MOIFEM from 1985 to 1990.

Gilah Leder is a professor of mathematics education in the Graduate School of Education at La Trobe University in Melbourne, Australia. Her teaching and research interests include gender issues, factors which affect mathematics learning, exceptionality, and assessment in mathematics. She serves on various editorial boards and educational and scientific committees and is currently president of the Australian Mathematical Sciences Council and of the Australian subcommittee of the International Commission on Mathematical Instruction.

Charlene Morrow is co-director of the SummerMath Program at Mount Holyoke College and a lecturer in psychology and education. She is a former president of Women and Mathematics Education, an affiliate group of the US National Council of Teachers of Mathematics. She holds a doctorate in clinical psychology from Florida State University and her current research interests include the investigation of women's approaches to learning mathematics, and cultural differences and similarities in mathematical attitudes among females. She is also an origami enthusiast.

James Morrow is co-director of the SummerMath Program at Mount Holyoke College and a lecturer in mathematics. He is the executive director of the NCTM affiliate, Women and Mathematics Education. His research interests include investigating the effects of a constructivist mathematics programme for students, and surveying attitudes of minority students toward mathematics. He holds a doctorate in mathematics from Florida State University.

Roberta Mura is a professor of mathematics education at Laval University in Québec, where she is also a member of the multidisciplinary Feminist Research group. She has a laurea in mathematics from the University of Milan and a doctorate, also in mathematics, from the University of Alberta. Between 1980 and 1984 she was the Canadian national coordinator of IOWME.

Cornelia Niederdrenk-Felgner studied mathematics and physics at the University of Tuebingen where she graduated with a doctoral degree. She then became a secondary-school teacher of mathematics and physics, and still works as a co-author of mathematics school textbooks. Currently, she works at the German Institute for Distance Education providing in-service training for mathematics teachers, including coordinating the 'Girls and Computers' project.

Pat Rogers is a professor in mathematics and education at York University in Toronto. She has a doctorate in mathematics from the University of London in the area of mathematical logic and is currently working in the area of curriculum reform and effective mathematics pedagogy at the undergraduate level. For the past five years, she served as founding academic director of the Centre for the Support of Teaching at York University. She was Canadian national coordinator of IOWME from 1984 to 1988 and co-organized, with Gabriele Kaiser, the IOWME sessions at ICME-7 in Québec. Educated in England, she now lives in Toronto with her daughter, Kate.

Nancy Shelley has spent a life-time in education. She pioneered the work on women and mathematics in Australia and was the founding Convenor of IOWME, holding that office from 1976–84. She has presented papers in the IOWME sessions of ICME-4 in Berkeley — that is in the published record of that Congress — and at ICME-5 in Adelaide.

Claudie Solar is a professor of mathematics education in the Faculty of Education at the University of Ottawa. She holds a masters degree in mathematical optimization and a doctorate in education. Her fields of interest are feminism and education related to both children and adults, mathematics education and technology. She is currently doing research in the area of equity and group learning, inclusive pedagogy, and on the use of calculators in elementary schools. She has been president of MOIFEM since 1992.

Neela Sukthankar has been living in Papua New Guinea for the past fifteen years and works in the Department of Mathematics and Computer Science, at the Papua New Guinea University of Technology. She has a doctorate in mathematics from the Tata Institute of Fundamental Research, Bombay, and her research interests include gender and education in Papua New Guinea. She is the founder editor of the Papua New Guinea Journal of Mathematics, the founding organizer of the nationwide Papua New Guinea Mathematics Competition, and the national coordinator for Papua New Guinean section of IOWME.

Denisse R. Thompson holds a PhD in education from the University of Chicago. She currently has a position in mathematics education at the University of South Florida with responsibility for preparing prospective teachers and updating practising teachers. Her research interests include curriculum development, the impact of technology on curriculum, issues of assessment, the study of proof, and the effective preparation of prospective mathematics teachers.

Sue Willis is a professor of mathematics education at Murdoch University in Western Australia. Her research interests are in gender and education and informed numeracy, the latter also from a social justice perspective. Her research and curriculum-development interests are linked by a commitment to mathematics for social justice.

Marjolijn Witte is a researcher at the Centre for Innovation and Cooperative Technology of the University of Amsterdam. She is interested in the design of educational environments open to all learners and is finishing her doctorate in the area of girls and mathematics education.

Index